Data Science and Big Data Analytics in Smart Environments

Editors

Marta Chinnici
ENEA - Italian National Agency for New Technologies
Energy and Sustainable Economic Development
Energy Technologies and Renewable Sources Department–ICT Division
Rome, Italy

Florin Pop
University Politehnica of Bucharest
Computer Science Department
Bucharest, Romania

Cătălin Negru
University Politehnica of Bucharest
Computer Science Department
Bucharest, Romania

CRC Press
Taylor & Francis Group
Boca Raton London New York

CRC Press is an imprint of the
Taylor & Francis Group, an **informa** business

A SCIENCE PUBLISHERS BOOK

First edition published 2021
by CRC Press
6000 Broken Sound Parkway NW, Suite 300, Boca Raton, FL 33487-2742

and by CRC Press
2 Park Square, Milton Park, Abingdon, Oxon, OX14 4RN

© 2021 Taylor & Francis Group, LLC

CRC Press is an imprint of Taylor & Francis Group, LLC

Library of Congress Cataloging-in-Publication Data
Names: Chinnici, Marta, editor. | Pop, Florin, editor. | Negru, Cătălin,
    1983- editor.
Title: Data science and big data analytics in smart environments / editors,
    Marta Chinnici, senior researcher, ENEA--Italian National Agency for New
    Technologies, Energy and Sustainable Economic Development, Roma, Italy,
    Florin Pop, Computer Science Department, University Politehnica of
    Bucharest, National Institute for Research and Development in
    Informatics, Bucharest, Romania, Cătălin Negru, Computer Science
    Department, University, Politehnica of Bucharest, Romania.
Description: First edition. | Boca Raton : CRC Press, 2021. | Includes
    bibliographical references and index. | Summary: "Many applications
    generate large datasets, like social networking and social influence
    programs, smart cities applications, smart house environments, Cloud
    applications, public web sites, scientific experiments and simulations,
    data warehouse, monitoring platforms, and e-government services. This
    book will primarily encompass practical approaches that advance research
    in all aspects of data processing, data analytics, data processing in
    Cloud/Edge/Fog systems, having a large variety of tools and software to
    manage them. The book focuses on focuses on topics covering algorithms,
    architectures, management models, high performance computing techniques
    and large-scale distributed systems"-- Provided by publisher.
Identifiers: LCCN 2020055953 | ISBN 9780367407131 (hbk)
Subjects: LCSH: Smart cities. | Big data. | Quantitative research.
Classification: LCC TD159.4 .D393 2021 | DDC 307.760285/57--dc23
LC record available at https://lccn.loc.gov/2020055953

ISBN: 978-0-367-40713-1 (hbk)
ISBN: 978-0-367-77603-9 (pbk)
ISBN: 978-0-367-81439-7 (ebk)

Typeset in Times New Roman
by Radiant Productions

Preface

Many applications like social networking and social influence programs, smart cities applications, smart house environments, cloud applications, public web sites, scientific experiments and simulations, data warehouses, monitoring platforms, and e-government services generate massive datasets. This book will primarily encompass practical approaches that advance research in Data Science and related applications involving Big Data and on all aspects of the Data Science World such as Data Analytics and Data Processing in different types of systems: Data Centers, Cluster Computing, Grid Computing, Peer-to-Peer, Cloud/Edge/Fog Computing, all involving elements of heterogeneity, and having a large variety of tools and software to manage them. The central role of resource management techniques in this domain is to create suitable frameworks for development of applications and deployment in smart environments, concerning high performance. The book also aims to focus on topics covering algorithms, architectures, management models, high-performance computing techniques and large-scale distributed systems with references to energetic sustainability.

Chapter 1 This chapter shows how to accelerate Domain Decomposition (DA) simulations introducing Machine Learning models and methods in the DA process at the top layer of the math stack. Experimental results are provided for pollutant dispersion within an urban environment. DA is the approximation of the true state of some physical system at a given time by combining time-distributed observations with a dynamic model in an optimal way.

Chapter 2 This chapter shows the problem of Edge Data Centers (EDCs) placement by presenting mobility-aware solutions for edge data center deployment in urban environments. The objective of Multi-access Edge Computing (MEC) is to better support low-latency applications by bringing storage and computational capabilities at the edge of the network into EDCs. Effective placement of EDCs in urban environments is key for proper load balancing, outage minimization and energy efficiency. This chapter takes into account the mobility of citizens and their spatial patterns to estimate the optimal placement of MEC EDCs in urban environments that minimize outages and energy efficiency.

Chapter 3 This chapter presents the design and implementation of a semantically enhanced data model for representing and analyzing the energy characteristics of a data centre (DC) as well as the reasoning-based methods used to assess the energy efficiency of the DC operation for optimal smart energy grid integration. Moreover, it shows how the model can be used to support data-driven decision-making operations to increase the energy efficiency of a test case.

Chapter 4 This chapter proposes a framework named Occupant Behaviors in Dynamic Environments (OBIDE) which is used to construct a system for monitoring the occupancy behaviors in a building using historicized spatio-temporal data and semantically enriched building information. The proposed system provides a centralized knowledge base to study occupancy behaviors not only for safety monitoring but also for different facility management applications that deal with dynamic building environments.

Chapter 5 This chapter proposes a new Belief Rule-Based Adaptive Particle Swarm Optimization (BRBAPSO) where the tuning parameters are adjusted dynamically by considering uncertainties, which ensure a balance between exploitation and exploration in the search space.

Chapter 6 This chapter presents an overview of NoSQL Time Series databases and Big Data machine learning techniques and analyzes their strengths and weaknesses to make an informed decision when choosing the best solution for processing, storing, managing, and analyzing Time Series.

Chapter 7 This chapter presents the problem of Spatial Decision Support Systems (SDSS), which are typically designed to face a specific issue by integrating features from a given theme, thus keeping the various research domains isolated and making the intersection among them difficult to be exploited where spatial data science may be a solution. This goal, namely translating data into insight, represents the preliminary phase of the challenge that researchers and analysts have set to obtain value from territorial knowledge.

Chapter 8 This chapter discusses research issues related to applications for energy performant buildings and Smart Cities. Buildings are responsible for 40% of the EU's total energy consumption and 36% of CO_2 emissions, making them one of the biggest contributors to global warming emissions in Europe.

Chapter 9 This chapter aims to analyze a dataset of plants and configures the most suitable plant species for a particular area, hoping that research in this area will elevate the planting process by reducing time and effort involved while deciding the suitable plant type for a given area. The method carried out for analysis is uses a dataset that consists of 104,352 plant species planted in the city of Greater Geelong council region, Australia.

Chapter 10 This chapter introduces a vehicular security verification and validation model. The model is an ontology-based approach that aims to create a knowledge representation of the vehicular components, assets, threats, and other aspects with all related data. Then it performs verification and validation to determine whether or not the security requirements are met under the actual conditions.

Chapter 11 This chapter discusses the usage of dynamic resource provisioning methods using cognitive intelligent networks based on the stochastic Markov decision process. One of the major challenges in the near future will be the ability to identify appropriate methodologies for the transmission of information on the channel's status and on the active reallocation of system resources among the operating nodes of the network.

Chapter 12 This chapter discusses related issues about water management and the problem of efficiently planning the utilization of water resources. Water management data is scattered across many platforms, providers or websites, which makes the task of efficiently planning the utilization of its resources difficult for managers and authorities.

Acknowledgment

The guest editors would like to thank all authors for their willingness to work on these complex book projects. We thank the authors for their interesting proposals of the book chapters, their time, efforts and their research results, which makes this volume an interesting complete monograph of the latest research advances and technology development on data science and big data analytics in smart environments. We also would Our special thanks go to the editorial and managerial team from Taylor& Francis for their assistance and excellent cooperative collaboration to produce this valuable scientific work. Finally, we would like to send our warmest gratitude message to our friends and families for their patience, love, and support in the preparation of this volume. We strongly believe that this book ought to serve as a reference for students, researchers, and industry practitioners interested or currently working in Data Science context and the Big Data domain.

Contents

Chapter 1

Mobility-Aware Solutions for Edge Data Center Deployment in Urban Environments

Piergiorgio Vitello,[a], Andrea Capponi,[a] Claudio Fiandrino[b] and Guido Cantelmo[c]*

1.1 Introduction

The fifth generation (5G) of cellular mobile networks is tailored to support high mobility, massive connectivity, extremely high data rates and ultra low latency. For this, it relies on Software-Defined Networking (SDN) and Network Function Virtualization (NFV). Radio access and core functions are virtualized and executed in edge data centers (EDCs) according to the Multi-access Edge Computing (MEC) principle. MEC was formerly known as Mobile Edge Computing and is standardized by the European Telecommunications Standards Institute (ETSI) [137]. MEC provides computing service closer to the end user, and thus finds applicability in scenarios where locality and low-latency are essen-

[a] University of Luxembourg.

[b] IMDEA Networks Institute.

[c] Technical University of Munich.

* Corresponding author: piergiorgio.vitello@uni.lu

tial[1] [343]. MEC is not tied to a single radio technology, but embraces both cellular and other radio access technologies such as WiFi and is agnostic to the evolution of the mobile network itself (4G or 5G).

The *edge*, also known as MEC host, is a (nano) data center deployed within the mobile network operator (MNO) infrastructure. The edge provides computing functionalities, can aggregate virtualized core and radio network functions and enables resource-constrained mobile devices to prolong battery lifetime while enhancing and augmenting performance of the mobile applications [119]. To this date, the research on edge computing has mainly focused on resource management and allocation by trading power consumption and communication delays [94, 381] while seminal works have mainly focused on the definition of architectural design principles [308, 261]. Emulation platforms for research in the area have only started to appear recently [122] and little attention has been paid to the problem of EDC deployment.

The EDC deployment is a particularly interesting and challenging problem in the context of smart cities. In spite of this, only a vision paper has explored such areas by assessing the feasibility of leveraging three different infrastructures, i.e., cellular base stations, routers and street lamps and analyzing the potential city coverage if only a subset of these elements was upgraded to furnish EDC capabilities [132]. This study is a step forward to solve the problems of coverage, EDC selection and user-to-EDC assignment. However, it does not consider urban mobility and the energy-cost of operating such infrastructure (e.g., to power on servers, base stations). Citizens mobility is influenced by many factors, including trip purposes and geographically imposed restrictions, such as home and work locations. These complex phenomena (herein called urban dynamics) influence the computational demand of BSs and EDCs that changes over time. A recent study unveils a strong correlation between the urbanization tissue, specifically land use, and the average traffic volume per-user by highlighting that different mobile services typically represent different temporal behaviors but uniform spatial patterns [129].

This chapter tackles the problem of EDCs deployment in a smart city context leveraging two observations. First, users access networks only with cellular connectivity and assume that EDCs should only be deployed at current Base Stations (BSs) sites to re-use already deployed infrastructure (e.g., power supply, cabinets on roofs). Thus, this solution is capital-expenditure free for MNOs. Second, this study focuses on human mobility. Within a city, complex dynamics regulate the inter-dependency of land use and citizen movements [179], i.e., the spatial distribution of citizens and locations they visit that determine mobility patterns. Similar to [397], this study exploits crowd sensed data (namely, Google Popular Times) to infer and predict human mobility with the goal of determining esti-

---

[1]In the rest of the chapter the words *user*, *mobile user*, and *citizen* are used interchangeably.

mates of computing demand and the optimal EDC deployment that minimizes outages. Third, this analysis targets energy-efficiency as one of the objectives of deployment. As it is well known from research in cloud computing [121] and radio access network [141], it is possible to reduce the energy expenditure in the two domains. For example, consolidating jobs/tasks into the minimum number of servers or active BSs in order to idle those with low workloads.

For the problem of EDCs deployment, this study proposes and compares three heuristics, namely distributed deployment algorithm (DDA), mobility-aware deployment algorithm (MDA) and Energy-aware MDA (EMDA) which enforce energy savings by powering off base stations with low utilization. The heuristics are evaluated with simulations that show how mobility-aware techniques outperform those that do not consider citizens' mobility. Simulators and emulators for fog computing/MEC are available. However, on the one hand, emulators provide a very fine precision of network and computing mechanisms that prevent assessing performance at city-level scale. On the other hand, simulators model either computing, networking aspect or both well while little support for realistic user mobility is enforced. This chapter specifically tackles this problem and proposes CrowdEdgeSim, the first simulator that simultaneously models urban mobility, network and computing dynamics. CrowdEdgeSim is key to assessing performance of MEC components, e.g., resource management techniques, at city level scale for smart environments. This chapter exposes the architecture of CrowdEdgeSim which features independent modules to model and simulate pedestrian mobility, mobile devices, network and computing resources and ETSI-compliant MEC applications. These three main factors influence each other in city-wide scenarios and are building blocks of the simulation environment, and require a further investigation on communications and MEC deployment. CrowdEdgeSim is based on the reference architecture of CrowdSenSim [120, 258], a simulator developed in the context of mobile crowd sensing [62] which exhibits scalability properties required for the purpose of this work.

1.2   Background and motivation

This section provides motivation of the work and presents background by first overviewing existing simulation and emulation tools for modeling and evaluating fog/edge computing systems and then it discusses how human mobility can impact the deployment of edge resources in urban environments.

1.2.1   Related works

MEC allows resource-constrained mobile devices to offload computational workload to nearby EDCs. For the details of fog/edge architectures and protocols, the interested reader can refer to existing surveys [261, 343]. Paradrop provides computing and storage capabilities and allows third parties to develop new types of services [224]. It includes a flexible hosting substrate in the WiFi APs that support multi-tenancy, a cloud-based backend, and an API for third-party developers. The integration of lightweight virtualization (LV) with edge is used for three different cases namely, i.e., autonomous vehicles, smart city, and augmented reality, which are discussed in [259].

The human-driven edge computing (HEC) paradigm leverages the mutual benefits of MEC and mobile crowd sensing (MCS) [48]. Some approaches propose a resource allocation model at the edge [381] and a multi-cloudlet infrastructure in the context of smart city scenarios [132]. The closest work to ours is [182], where the authors present an optimal cloudlet placement and allocation of users to them in Wireless Metropolitan Area Networks (WMANs). Nevertheless, cloudlets placement does not capture user movements within the network at a scale. These factors are important as nowadays smart cities are characterized by increasing mobility that influences the workload and computational demand of edge resources.

The study presented in this chapter fills this gap as it quantifies the importance of user mobility at the city scale to estimate the computational demand and deploys EDC accordingly.

1.2.2 Simulation and emulation platforms for fog and edge environments

Simulators

iFogSim focuses on fog resource management [145] and allows the measurement and evaluation of latency, network congestion, energy consumption, and costs of different policies to efficiently deploy resources in IoT environments at the edge of the network. Specifically, the core component deals with resource management, i.e., resource deployment and scheduling. *FogTorch II* is developed in Java and allows the estimation of the QoS-assurance of composite Fog applications deployments based on probability distributions of bandwidth and latency of communication links (through Monte Carlo simulations). To this end, it considers constraints due to both processing (for example, CPU, RAM, storage, and software) and QoS (e.g., latency, bandwidth).

The above tools unfortunately do not consider human mobility, which is the core of this work. Interestingly, the authors of *iFogSim* discuss the support of mobility as the first step of future directions to extend their simulator. Edge-CloudSim [332] and FogNetSim++ [288] provide limited support to mobility but do not consider how realistic human mobility in urban environments can impact the deployment of edge resources.

Emulators

Many emulation platforms have been recently proposed for fog/edge computing. openLEON [122] is an open-source platform that enables experimentation and prototyping in multi-access edge computing. openLEON can emulate end-to-end activities, from the edge data center (modeled with the popular Mininet) to the wireless-connected end user (the mobile network is emulated with software defined radio srsLTE software). EmuFog [245] enables the researchers and developers to create fog computing infrastructures and incorporates network topologies into MaxiNet [376]. However, MaxiNet does not support wireless communications. MockFog [157] captures well the computing dynamics because it hosts the entire emulation process of applications over computing, storage and memory resources in the cloud. In other words, MockFog emulates fog architectures consisting of edge devices such as Raspberry Pis, cloudlets and clouds over infrastructures like Amazon EC2. The main shortcoming of such an approach is the poor emulation of network dynamics as no notion of protocols nor technology is given (only simple parameters like incoming/outgoing maximum rates, delay, losses, reordering time, duplicate packets are captured). Unlike MockFog, EmuEdge [392] significantly strengthens the precision of network emulation that is based on Linux network namespaces. With Xen, EmuEdge supports both containers and virtual machines.

1.2.3 The impact of human mobility and scalability in realistic urban environments

The scalability of existing solutions to city-wide scenarios is crucial. Scalability at city-level scale means taking into account the number of citizens that use mobile devices (e.g., while commuting during a working day or going out at weekends). Decades of research in travel behaviour showed that human mobility is often repetitive [138], thus contrasting the typically employed random models. Traffic simulators like SUMO require a large amount of data for proper calibration. This data includes but is not limited to, travel surveys (i.e., user reporting their trip data), socio-demographic characteristics of the population, geometric design of roads, and traffic data. Without proper calibration, traffic models are likely to produce biased mobility patterns and wrong traffic estimations. Cellular traces have been used to partially overcome this issue. For example, it unveiled that the Expo Area in Milan during 2015 was very active while the information available from the municipality was reporting otherwise [128]. Cellular network traces can not only predict the impact of user mobility on traffic management during regular days and on those having special events [331], but can also unveil specific characteristics, such as the diversity of baseline communication activities across countries [129].

CrowdEdgeSim simulator was developed to fill these gaps. The CrowdEdgeSim built-in mobility model can reproduce simplified mobility patterns when limited information is available. Given the modular nature of CrowdEdgeSim, when a more elaborate mobility model becomes available - such as SUMO - user trajectories can be included seamlessly without having to modify the code significantly. Similarly, real data (anatomized user position for example) can also be used.

1.3 Modeling human mobility in urban environments

Citizens movements and their social interactions influence the demand for computing resources. Different models can be exploited to generate pedestrian mobility in urban environments. This work proposes three different models according to user arrivals. In the first model, the user arrivals are distributed according to a probability distribution (e.g., uniform, gaussian) over the simulation period (P-MOB). The second mobility model derives user arrivals from a pedestrian dataset (D-MOB). The third model estimates user arrivals directly from the Point of Interest (POI) attractiveness data (A-MOB).

1.3.1 The P-MOB mobility model

With P-MOB, user arrivals are generated during the simulation period according to a specific uniform probability distribution, for example, in such a case, the average total number of users that start walking over a certain time window is constant. Another distribution is when user arrivals are randomly distributed with a Gaussian probability with a certain peak value. A representative example is given by a work that focuses on smart lighting [60], in which the authors generate mobility according to a binomial distribution that has a lower concentration of citizens walking at night.

1.3.2 The D-MOB mobility model

The D-MOB models mobility by using real traces from different available datasets on human mobility. For instance, one of the most popular datasets for MCS activities is ParticipAct, originating from a MCS campaign of around 170 students in the Emilia Romagna region (Italy) [73]. Without having a dataset at their disposal, in a former work the authors extracted the profile of the average number of contacts in 7 days and used it as a reference to determine the user arrivals [365]. Given the total simulation period in days, the authors subdivided the period into hours and they estimate the minimum number of users to be allocated so that the average user contact follows the ParticipAct profile. Note that a user contact occurs when two users are within a certain distance R. For simplicity, they count unique contacts even when users trajectories intersect multiple times. Additionally, if a calibrated traffic model is available, D-MOB can leverage these trajectories to approximate real traces.

1.4 The A-MOB mobility model

A-MOB mobility model is based on the attractiveness of POIs and local businesses (LBs). This depends on the social interactions of citizens with urban environments and varies with the time and day of the week according to several factors, such as commuting for work or particular events. This mobility could be generated from several datasets (e.g., social networks like Foursquare). This study leverages Google Popular Times to estimate citizens mobility that reflects daily urban patterns and allows the investigation of how spatial aspects and social interactions of citizens influence the computational demand (workload) of the edge. Google Popular Times define the popularity of a LB with per-hour values normalized between the weekly maximum and minimum number of customers in each. However, the real number of customers remains unknown. Hence, one of the contributions of this study explained hereafter, is a new approach to estimate the number of customers from the coarse measurements available. The popularity metric is then used to approximate users' temporal distribution among different LBs.

For each type of LB t, a random value of the typology "t" is generated between 0 and N_{max}^t, where N_{max}^t is the maximum value of customers. This value and the average waiting and staying time in a LB permits the computation of the number of people remaining with it. The aggregation of different LBs in a region defines how crowded a district is. In summary:

■ For each local business L of typology t the maximum number of people N_L^t visiting that specific LB is drawn from the uniform distribution $[0; N_{max}^t]$.

■ The time-dependent demand for a certain L is obtained by combining the number of visits in popular times:

$$D_{L,h} = P_{h,L} \cdot N_L^t, \tag{1.1}$$

where $D_{L,h}$ is the number of users visiting location L during time interval h and $P_{h,L}$ is the popularity index according to the exploited data.

■ Finally, for a certain geographical area d, the overall computational demand is obtained as the combination of all LBs located in that specific area as:

$$A_{d,h} = \sum_{l \in L_d} D_{l,h}, \tag{1.2}$$

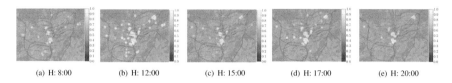

(a) H: 8:00   (b) H: 12:00   (c) H: 15:00   (d) H: 17:00   (e) H: 20:00

Figure 1.1: Traffic generation in Luxembourg City at different hours of a working day following A-MOB.

where $A_{d,h}$ is the demand of a certain geographical area d at time instant h and L_d is the sub set of LB located in d.

Some preliminary results to expose how the demand varies in a city during the day is presented in Fig. 1.1. It shows that the heatmaps of potential demand of the BSs in Luxembourg City, following the A-MOB mobility based on Google Popular Times[2] in a weekday. BSs under heavy loads are around the railway stations when citizens commute (i.e., H: 8:00 and H: 17:00), in the city center during lunch and dinner time, and in the university area during the day. By considering different hours of a day, loads of BSs vary and, in turn, the potential computational demand at EDCs.

1.5 Modeling MEC scenario for data center deployment in urban environments

1.5.1 Modeling MEC components

This subsection describes the MEC model exploited for this work. The first part models the computational and communication aspects of the MEC scenario, the second part discusses the parameters used to evaluate the performance of the system.

The computational demand depends on the aggregated traffic generated by the tasks that the user application initiates. Each EDC can sustain a computational demand that is generated by the users which are currently connected to the set of BSs that the EDC is managing. MEC resources define the capability of the system in accomplishing the tasks requested. Specifically, this study considers both the number of data centers distributed around a city scenario, and the resources within each data center, such as the number of servers and their performance.

The ratio between amount of received computational demand and the amount of resources required to satisfy it defines the efficiency of a MEC system. When no resources are available, some of the demand can not be satisfied and defined as an outage. Outages are the main key performance indicators (KPIs). To access the

---

[2]https://support.google.com/business/answer/2721884.

EDC processing capacity, a user sends a request message which is acknowledged when a sufficient amount of resources available with it permit its accommodation. The latency L of the request/reply exchange includes a *network delay* and a *processing delay* [123]. The network delay D_p consists of different components, such as transmission, propagation, queuing, and routing. The processing delay D_c depends on the network processing packet in a NFV environment. The latency L is given by:

$$L^i = D_p^{i,k} + D_c^k + D_p^{k,i}, \qquad (1.3)$$

where $D_p^{i,k}$ is the *network delay* from user i to EDC k, D_c^k is the *processing delay* at EDC k, and $D_p^{k,i}$ is the *network delay* from EDC k to user i. $D_p^{i,k}$ is given by the time required to transmit a single packet from user i to EDC k through the BS b:

$$D_p^{i,k} = T_{wireless}^{i,b} + T_{wired}^{b,k}, \qquad (1.4)$$

and viceversa for $D_p^{k,i}$ which is the reverse path. $T_{wireless}^{i,b}$ is the time required for transmission from user i to BS b through the LTE interface, and $T_{wired}^{b,k}$ is the transmission time from a BS b to EDC k. $T_{wireless}$ is given by:

$$T_{wireless} = \frac{S}{\tau_i}, \qquad (1.5)$$

where S is the size of the packet and τ_i is the transmission rate of user i. τ_i is obtained by computing the channel quality indicator (CQI) and in turn the modulation and coding scheme (MCS) used by the LTE scheduler as explained in 1.7 and 1.15.1 respectively. T_{wired} is given by the sum of the link propagation and transmission delays as explained in eq. 15 of [121].

1.5.1.1 Problem formulation

This subsection formulates the problem of EDC distribution in urban environments. Let $\mathcal{B} = \{b_1, ..., b_{N_B}\}$ be the set of BSs, each characterized by a latitude and longitude that defines the location of b_i in the city the set of EDCs to be deployed, $\mathcal{E} = \{e_1, ..., e_{N_E}\}$ where $N_E < N_B$ and $\mathcal{E} \subset \mathcal{B}$. EDCs are deployed at BS sites to re-use already deployed like power supply and cabinets on the roof, for example (e.g., power supply, cabinets on roofs). This provides a full spatial coverage and is more cost-effective than creating new EDC sites [132]. Similar to [218], this work considers the latency outage probability of the system O as the Key Performance Indicator (KPI), defined as:

$$O = Pr\{L \geq D_{max}\}, \qquad (1.6)$$

where L is the latency the user observes and D_{max} is the maximum acceptable delay bound defined in the form of a SLA agreement for the current application. While in [218] the delay outage probability indicates the one-way latency, in this study the *latency outage probability of the system* needs to measure the Round-Trip-Time (RTT) to capture the fact that if a user does not receive a reply from the EDC, the task is not accomplished and O increases. This can happen for two reasons:

■ the EDC rejects the incoming task because it is overloaded;

■ the user does not receive the reply in due time because of either processing or networking delays.

Given a fixed number of EDCs N_e to deploy, the problem consists of finding a match with existing BS locations to minimize the average latency outage probability of the entire system to maximize the computational capacity of the system, for example (e.g., to maximize the computational capacity of the system):

$$\min_{E} O. \tag{1.7}$$

where E is the set of BSs that are chosen to place the EDCs.

1.5.2 Accounting for energy consumption

This subsection presents a model to compute the energy consumption. It includes energy costs incurred for transferring data to and computing processing costs within the EDC using network communication technologies.

1.5.2.1 Network components

To model the energy consumption attributed to the network components like BSs, for example this work refers to [230], which correlates the BS power consumption (in W) with the traffic load mentioned as in Erlang (see Fig. 26 of [230]).

1.5.2.2 EDCs

Previous study on energy-consumption of data centers highlights a non-linear relationship between load and energy consumption. For example, activating a new server in an already operational rack does not require additional activated network switches [121]. The power consumption $P(l)$ of a server is computed assuming it implements Dynamic Voltage and Frequency Scaling (DVFS)[3] as follows:

$$P(l) = P_{\text{idle}} + \frac{P_{\text{peak}} - P_{\text{idle}}}{2} \cdot \left(1 + l - e^{-\left(\frac{l}{\tau}\right)}\right), \tag{1.8}$$

where l is the load of the server and τ is the utilization level at which servers attain asymptotic power consumption, which is typically in the range $[0.5, 0.8]$.

---

[3]Available on: http://www.spec.org/power_ssj2008/.

1.6 EDCs deployment policies

The MEC deployment operation consists of deploying EDCs at the edge of the network, close to end users. This work specifically targets deployments in urban environments. For cloud data centers, deployment strategies are driven by economical reasons (for example, the cost of energy to power the facility and cooling it down). Large companies like Facebook and Apple are exemplary in deploying data centers near cheap hydro-power sources [65]. Conversely, in edge computing, the objective of the deployment is to effectively support low latency applications. To this end, it is necessary to deploy EDCs in a way that supports the computational demand in accordance with time and spatial patterns of citizen movements.

The placement policy and BSs assigned to EDCs are keys for effective EDC deployment. To deploy edge computing capabilities in urban environments aiming to minimize the outage probability with fixed resources (or to minimize resources given a fixed target outage probability) two fundamental aspects must be considered. First, the placement planning of EDCs around the city and how to allocate BSs to different EDCs.

This work proposes and compares three different policies. The first policy is, *distributed deployment algorithm (DDA)*, which deploys EDCs such that they are the centroids of a cluster composed of a set of base stations which share a similar distance. The second policy is, *mobility-aware deployment algorithm (MDA)*, which considers the mobility of citizens and their social interactions in urban environments to calculate the expected computational demand and distribute edge resources by utilizing them as weights. The third policy is *energy-efficient mobility-aware deployment algorithm (EMDA)* enhances MDA's ability to save energy by switching off some BSs given the placement of MDA in EDCs.

The following subsection explains the details of the three policies. By comparing such schemes, this work asserts that considering the spatial and temporal distribution of citizens significantly improves MEC performance and permits the saving of energy while maintaining the same MEC outage performance.

1.6.1 Distributed Deployment Algorithm (DDA)

The Distributed Deployment Algorithm (DDA) distributes EDCs according to the k-medoids clustering algorithm. DDA places EDCs and assigns BSs to them by exploiting the k-medoids clustering algorithm, which is similar to the more famous k-means. While in the k-means algorithm the centers of clusters are not necessarily input data points, this method chooses the centroids in between the input data. In other words, it clusters BSs and chooses EDCs as centers of the clusters in between the BSs within a cluster, which perfectly fits the purpose of this study. More specifically, DDA assigns EDCs among BSs by computing a cost based on the distances between BS candidates and EDCs. The main short-comings of this approach consist in some under-utilized EDCs and others that suffer from big delays due to overloads of computational demand leading to high values of outage probability, as discussed more in details in 1.14. To overcome these issues, two possible directions can be investigated. First, to propose a more effective placement of EDCs among the BSs (see *MDA* section later on). Second, to allocate servers among EDCs in proportion with the computational demand (see *Allocation of Servers among EDCs* section later on).

1.6.2 Mobility-aware Deployment Algorithm (MDA)

MDA aims to decrease the overall outage probability of the system by consider-ing where the computational demand is higher according to the spatial patterns of citizens. In other words, the MDA solution is based on the idea to consider the complex dynamics of a city (e.g., user mobility and social interactions) to pro-pose a more effective placement of edge resources. Like the DDA, MDA places EDCs among BSs by exploiting the k-medoids algorithm. As opposed to DDA, it assigns EDCs among BSs by computing a cost based on the number of requests received by them and the corresponding computational demand for EDCs. MDA is based on an iterative approach (see the MDA pseudocode in Algorithm 1) that computes how far the total number of requests for each EDC is from the aver-age number of requests each EDC should have. Specifically, note that line 28 calculates the previously explained costs.

1.6.3 Energy-efficient Mobility-aware Deployment Algorithm (EMDA)

EMDA maintains the same EDC deployment algorithm proposed in MDA. Fur-thermore, it tailors energy-savings by powering off BSs strategically. Specifi-cally, a BS is powered off for a duration W when its load compared to the previ-ous window $W - 1$ is below a given threshold of the total BS load. After having been off during W, it is activated again in $W + 1$. Obviously, when off, no traffic can flow through a certain BS and this traffic is re-routed to other active ones, thereby providing consolidation.

1.6.4 Allocation of servers among EDCs

Besides the deployment itself, the decision on the number of servers to be allo-cated per EDC is crucial. This issue is described by formulating the following problem: given a certain fixed number of servers, how should they be allocated among EDCs? To this end, this study compares a fixed number of servers (FNS) approach and a proportional number of servers (PNS) approach. The former sim-

---

Algorithm 1 MDA

1: **Input:** B, R, n ▷ BSs, requests at BSs, number of EDCs
2: **Output:** E, C ▷ Set of BSs chosen as EDCs, Set of BSs-EDCs connections
3: $AVG\_reqs = sum(R)/n$ ▷ Average num. of requests per EDC
4: $E \leftarrow Random(B, n)$ ▷ Randomly select n BSs as EDC
5: $Cost \leftarrow getCost(E, AVG\_reqs, B, R)$ ▷ Compute cost
6: **while** *Swaped* **do**
7:     $Swaped \leftarrow False$
8:     **for** $b \in B$ **do**
9:        **for** $i = 1 \rightarrow n$ **do**
10:           $E\_dup = E$ ▷ Copy of E
11:           $E\_dup_i \leftarrow b$
12:           $Cost\_dup = getCost(E\_dup, AVG\_reqs, B, R)$
13:           **if** $Cost\_dup < Cost$ **then**
14:              $E \leftarrow E\_dup$
15:              $Swaped \leftarrow True$
16:           **end if**
17:        **end for**
18:     **end for**
19: **end while**
20: **procedure** GETCOST(E, AVG_reqs, B, R)
21:     $Cost = 0, R\_tot = \emptyset$
22:     **for** $b \in B$ **do**
23:        $e = ClosestEDC(b, E)$ ▷ Find EDC in E at minimum distance from b
24:        $C_b \leftarrow e$ ▷ Connect BS b to the nearest EDC e
25:        $R\_tot_e \leftarrow R\_tot_e + R_b$ ▷ Sum requests at EDC
26:     **end for**
27:     **for** $e \in E$ **do**
28:        $Cost \leftarrow Cost + |R\_tot_e - AVG\_reqs|$ ▷ Compute cost
29:     **end for**
30:     **return** Cost
31: **end procedure**

---

ply allocates the same number of servers in each EDC. The latter distributes the total number of servers in proportion with the computational demand of each EDC (taking into account its temporal evolution).

1.7 The simulator: CrowdEdgeSim

This work presents CrowdEdgeSim, a simulation tool developed for MEC scenarios. The Simulator is able to simultaneously model network and computing dynamics and urban mobility. This section introduces the hierarchical architecture of the simulator, as shown in Fig. 1.2. It is composed of three main independent modules that describe *urban mobility*, *MEC scenario*, and *mobile devices*. These three modules influence two other modules, that is, *MEC deployment* and

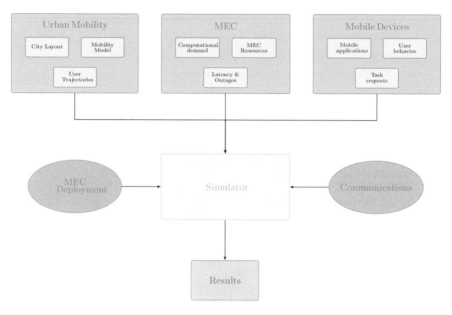

Figure 1.2: CrowdEdgeSim architecture.

Communications. All the five modules are inputs to the simulator engine, which provides the output results.

1.8 MEC scenario

This module models the computational aspects of the MEC scenario and the considered parameters to evaluate the performance of the system. The implementation of these MEC components follows the model widely described in Section 1.5.1.

1.9 Urban mobility

To approximate and simulate urban mobility, the simulator considers the city layout, the mobility model and user trajectories. The city layout consists of the nodes forming the street networks of a city where citizens can move. The coordinates of the nodes forming the street networks are obtained by applying an OSM procedure that augments the precision of the network resolution (AOP) [365]. OSM provides the graph of the street network $G_{OSM} = (V, E)$, where V is the set of vertices or nodes and E the set of edges. The objective of AOP is to increase the precision of G_{OSM} to achieve a given *target distance* D between any two adjacent nodes. Figure 1.3 compares two different versions of the street network graph of Luxembourg city. It highlights the different granularity of the street net-

(a) Luxembourg - Precision of OSMnx   (b) Luxembourg - Precision of AOP

Figure 1.3: Different granularity for Luxembourg City. Red circles denote the vertices provided by OSM and green circles those created by AOP.

work graph obtained with OSM (see Fig. 1.3(a)) and AOP (see Fig. 1.3(b)). AOP is general enough to be applied to any situation where the original precision of OSM is not sufficient.

The mobility model defines how citizens move on the street networks (for example, random walking, uniformly distributed patterns or datasets). The simulator supports different models (see Section 1.3).

The user trajectories consist of a starting point, and ending point and the exact path in between and along the street networks, under the constraint of the walking time period. As shown in Algorithm 2, the paths strictly depend on both the chosen city scenario and the mobility model, and the inputs to compute a user trajectory are the street network graph and the area demand probabilities, where the first is provided by the city scenario while the second comes from the mobility model (e.g., eq. 1.2 defines the area demand probabilities of A-MOB). The procedure to define a user trajectory starts by selecting the area represented by the origin exploiting the area demand probabilities (line 3). Then the origin is randomly picked between the nodes of the selected area subgraph. All the paths starting from the point representing the origin are generated (line 7) and from among them the final one within walkable distance is randomly selected.

1.10 Mobile devices

Nowadays, services that were typically run in the cloud are accessible in mobile devices (for example, gaming, video offloading, augmented reality). To be effective, mobile applications must respect some constraints (for example, low latency, wide bandwidth) that depend on network and computation capabilities and require resources at the edge of the network. To effectively deploy edge computing resources and study how they impact different aspects of mobile devices, like battery lifetime, this work considers different applications. These impact the utilization of mobile devices (for example, task requests, instructions per sec-

---

Algorithm 2 Trajectories Procedure

1: **Input:** P_h, G ▷ List of demand probability of different areas in hour h, Street network graph
2: **Output:** T_i ▷ Trajectory of user i
3: $D = Weighted\_Random\_Selection(P_h)$ ▷ Select area D based on the probability list
4: $G_D \leftarrow Split(G, D)$ ▷ Create a subgraph of area D
5: $N \leftarrow Nodes(G_D)$
6: $O = Random(N)$ ▷ Origin point randomly picked between subgraph nodes
7: $Paths \leftarrow AllPaths(G, O)$ ▷ All paths of graph G starting from node O
8: $w = WDist(i)$ ▷ Uses speed and walking time of user i to compute its walkable distance
9: $CPaths \leftarrow CutPaths(Paths, w)$ ▷ Cut paths to length w
10: $T_i = Random(CPaths)$ ▷ Randomly pick one of the cut paths
11: **return** T_i

---

ond, energy consumption) and pose different constraints. While delay-tolerant applications such as email or messaging have mild requirements, new applications require high-performance low-latency responses (for example, online gaming and augmented reality). To this end, this module characterizes different types of applications, their properties and constraints. Specifically they are:

- Synthetic models: Assign typical computing parameters and networking components to mobile applications (see later on this section for an exhaustive list);

- Real traces: They are based on measurements from experiments performed with the openLEON emulator [122].

Each mobile application is linked with typical parameters with certain specifications in the case under study. This subsection presents the general technical specifications considered for each application, while Section 1.15.1 discusses the values used for simulations. For each of these apps the following are considered:

- Usage percentage: the percentage of app utilization by users

- Task inter-arrival: the frequency of tasks sent to the EDC from mobile devices;

- Delay sensitivity: it describes how an application is tolerant to delays and is used to set the threshold of the maximum latency supported by an application;

- Active and idle periods: the time duration of periods in which the interface is active for transmitting or receiving data or idle while waiting for tasks to send or receive;

- Upload/download data: it consists of sent and received data for each app;

■ Task computational demand: it characterizes the computational demand for each task.

1.11 Communications

Urban mobility, MEC scenario, and mobile devices influence communications (e.g., the locations of antennas in a city, mobile applications that require specific constraints, and MEC resources that can or not guarantee low latencies). Nevertheless, the module of communications is independent of the others as, once the inter-dependent parameters are fixed, each communication technology has to be modeled alone (for example, the data rates, propagation delays that vary among garrotes connecting the EDCs and BSs - and wireless connections - from mobile devices to the BSs).

Users move in the street network and send task requests according to the mobile application and communication technology considered (for example, augmented reality or image processing) (for example, WiFi or LTE). Mobile devices connect to the closest available antenna, from a WiFi access point or a cellular base station respectively. The implementation of communication through WiFi interfaces is presented in a former work of the authors [355]. However, although available in the simulator, this work employs cellular connectivity and the communication model is implemented as follows. First, each smart device tries to connect to a BS, the association procedure between the user and the BS (see the pseudocode in Algorithm 3) gets initiated by checking if the device had a prior connection with a BS or not (new or disconnected). So, it tries to connect to the closest BS with an available slot for transmission (line 8), otherwise a handover to another BS takes place if the old one is overloaded (line 16) and the new one is not (line 22). Once the device is connected and transmission starts the simulator calculates the distance user-antenna and the channel quality indicator (CQI). CQI indicates the quality of the channel: in LTE it ranges between 1 and 15 and maps CQI to a modulation scheme defining the achievable throughput (for example, QPSK for CQI from 1 to 6, 16QAM for CQI from 7 to 9, and 64QAM for CQI from 10 to 15). CrowdEdgeSim calculates the end-to-end latency by computing the delays of the wireless communication from user to the connected BS through LTE, from the BS to the associated EDC by wired communications, and the computational time at the EDC to complete the requested task and for the reverse path back to the user.

1.12 MEC deployment

The MEC deployment operation consists of deploying edge data centers (EDCs) or micro data centers (MDCs) at the edge of the network, close to end users. This study specifically targets deployments in urban environments. For cloud

---

Algorithm 3 Association procedure between BS and Users

1: **Input:** B, i, old_bs_i ▷ List of all BSs, user i, BS of user i on previous timeslot
2: **Output:** bs_i ▷ BS on which user i is connected now
3: $P = Position(i)$
4: **if** isNull(old_bs_i) **then** ▷ Check if user was disconnected on previous timeslot
5:     **while** $Size(B) > 0$ **do** ▷ Check lenght of BSs list
6:         $b_i = ClosestBS(P,B)$ ▷ Find BS at minimum distance from user i
7:         $D = Distance(b_i, P)$
8:         **if** $D < MaxDistance$ & $Users(b_i) < MaxUsers_{b_i}$ **then**
9:             **return** b_i
10:         **else**
11:             $B \leftarrow Remove(b_i)$ ▷ Remove b_i from BSs list
12:
13:         **end if**
14:     **end while**
15: **else** ▷ user was already connected on previous timeslot
16:     **if** $Users(old\_bs_i) < UpperTHR$ **then**
17:         **return** $b_i = old\_bs_i$
18:     **else**
19:         **while** $Size(B) > 0$ **do** ▷ Check lenght of BSs list
20:             $b_i = ClosestBS(P,B)$ ▷ Find BS at minimum distance from user i
21:             $D = Distance(b_i, P)$
22:             **if** $D < MaxDistance$ & $Users(b_i) < LowerTHR$ **then**
23:                 **return** b_i
24:             **else**
25:                 $B \leftarrow Remove(b_i)$ ▷ Remove b_i from BSs list
26:             **end if**
27:         **end while**
28:     **end if**
29: **end if**

---

data centers the specific deployment strategies are driven by economical reasons (for example, the cost of energy versus power on the facility and cooling down). For example, large companies like Facebook and Apple have deployed data centers near cheap hydro-power sources [65]. Conversely, in edge computing, the objective is to deploy edge data centers to effectively support low latency user applications.

1.13 Simulator engine

The simulator engine is written in C++ and Python. Most parts on communications and generation of events are in C++, while the street network, city layout and mobility are coded in Python. The simulator takes an input off the five modules *MEC scenario*, *urban mobility*, *mobile device activities*, *communications*, and *MEC deployment* respectively and gives the resulting output according to the focus of the analysis.

1.14 Results

The results depend on the specific area of study and can be customized by researchers. For example, this chapter considers the total outage of the system and energy consumption as a key metric. The total outage of the system is given by the ratio of unaccomplished tasks to the total number of tasks in consideration.

1.15 Performance evaluation

The objectives of end-users and MNOs are often in conflict and lead to trade-offs. While end-users aim at accessing new services with low latency, high reliability and bandwidth, the MNO targets revenue maximization and resource usage minimization while preserving a certain degree of performance guarantees. This section evaluates the proposed EDC deployment heuristics.

1.15.1 *Evaluation settings*

1.16 City layout and mobility model

All the simulations resort to the center of Luxembourg City, which covers an area of 51.73 km^2 with a perimeter of 52.5 km and a population of 119214 inhabitants as of the end of 2018. It is home to many national and international buildings and headquarters of both financial and governmental institutions. It has a street network where users can move on pedestrian pathways having a granularity of 3 meters (i.e., the distance from one node to another one). The users in the simulations are 100000 pedestrians distributed on the street network according to A-MOB, using Google Popular Times consisting of 1083 LBs belonging

Table 1.1: Setup parameters.

DESCRIPTION	VALUE
Number of users	100000
Number of BSs	141
Number of EDCs	8
Number of servers per EDC	8
Server CPU speed	10 GIPS
Number of total LBs	1083
Number of LBs typologies	13

Table 1.2: LTE transmission rate.

MODULATION	τ_i
QPSK	15.84 Mbps
16QAM	30.576 Mbps
64QAM	75.376 Mbps
256QAM	97.896 Mbps

to 13 different categories like restaurants, pubs and public offices among others. The walking pedestrians are uniformly distributed between $[1,40]$ minutes and $[1,1.5]$ m/s for a simulation period of 24 hours.

1.17 MEC scenario and communications

The target is to deploy 8 EDCs around Luxembourg City each equipped with 8 servers unless specified differently. Each server supports 10 GIPS CPUs. EDCs are connected through optical fiber cables to 141 BSs which are placed according to their real locations in Luxembourg City[4]. The real coordinates (latitude and longitude) of BSs are taken from public mobile communication networks 50 W power[5].

The maximum number of connected users to a BS is uniformly distributed between [1,40]. The maximum number of possible users determines the maximum load of the BS, and thus the highest power consumption [230].

The transmission rate τ_i corresponding to user i is calculated as shown in Table 1.2, where the modulation scheme depends on the CQI as explained in 1.7.

Table 1.3: Applications parameters.

	SYNTHETIC MODELS			REAL TRACES
	AUGMENTED REALITY	HEALTH APP	COMPUTE INTENSIVE	VIDEO STREAMING
Usage Percentage (%)	30	20	20	30
Task Inter-arrival (sec)	2	3	20	2,6-3,7
Delay Sensitivity (%)	90	70	10	50
Active/Idle Period (sec)	40/20	45/90	60/120	120/0
Upload/Download Data (KB)	1500/25	20/1250	2500/200	3000/52
Task Computational Demand (MI)	9	3	45	0,1

[4] https://map.geoportail.lu/.
[5] https://data.public.lu/fr/datasets/cadastre-gsm/.

1.18   Mobile applications

The evaluation uses both synthetic traces taken from applications with known values and real traces obtained from experiments.

1.18.1   Synthetic applications

To generate traffic, this work considers three of the most popular mobile applications as typical examples for low-latency applications: compute intensive, augmented reality, and healthcare. Glassist uses augmented reality on Google Glasses and it is an application [328]. The compute intensive application generates tasks that require a large amount of CPU resources for their execution, but poses mild requirements to the network. The healthcare application offloads sensor data to the server for fall risk detection service which is then communicated to the person [358].

1.18.2   Application based on real traces

This work employed OpenLEON [122] to obtain real traces and utilizes an application to deliver video traffic over the Internet with Dynamic Adaptive Streaming over HTTP (DASH). By using DASH, a video is divided into chunks of different lengths that are encoded at multiple bit rates and stored with a Media Presentation Description (MPD) file. Wireshark is exploited to capture packet traces at the user and BS levels while performing experiments with Adaptive Bit Rate (ABR) video streaming. ABR is a technique that allows content providers to optimize video quality because it enables users to choose chunk resolution based on specific metrics dynamically. For example, the DASH client player estimates the available bandwidth and communicates with the server to specify the chunk representation to be downloaded among the ones indicated by the MDP file.

Figure 1.4 shows the openLEON setup. To implement the video streaming application and obtain real measurements, it has used a laptop equipped with an Intel i7-4600U processor at 2.1 GHz, 8 GB RAM and Linux Ubuntu 16.04 LTS to run the core network (srsEPC application from srsLTE version 18.3.1) and the data center network is emulated with Mininet [209]. A desktop computer equipped with an Intel i7-6700 processor running at 3.4 GHz, 16 GB RAM and Linux Ubuntu 16.04 LTS, runs the BS application (srsENB application from srsLTE version 18.3.1). The LTE physical BS is an Ettus B210 connected with USB 3.0 to the desktop computer. The UE as well uses an Ettus B210 connected to the laptop with an Intel i7-4600U processor upto a frequency of 2.1 GHz and 8 GB of RAM running the UE application (srsUE application from srsLTE version 18.3.1). The famous BigBuckBunny video was encoded with the ffmpeg tool by using multiple resolutions:

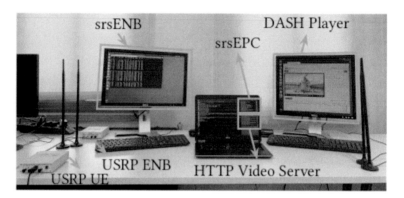

Figure 1.4: openLEON setup.

- 256 × 144 at a bit rate of 500 Kbps;

- 858 × 480 at a bit rate of 2800 Kbps;

- 1280 × 720 at a bit rate of 6000 Kbps.

The server side was configured by opening a Node.js http-server from a Mininet host and enabled the Cross-Origin Resource Sharing (CORS) option. At the client side, it was installed the DASH.js reference player and augmented its capabilities to give output statistics about video buffer levels and video bit rates.

Table 1.3 shows the setup parameters for the 4 different applications, divided into the synthetic models and the real traces obtained by openLEON.

1.18.3 Results

CrowdEdgeSim permits to generate different features of urban environments, MEC components, and smart device activities with corresponding generated traffic. Consequently, results depend on the case under study and the city layout, which can be customized by researchers according their needs. This study considers the end-to-end latency from/to the user, the outage of the system, and the energy consumption as KPIs.

1.19 Smart device activities and EDC deployment

Figure 1.5 shows the trajectories of 5 users moving on the street network of Luxembourg City. Circles represent the starting walking point and the star the ending point. Figure 1.5(a) shows the generated usage of mobile phones according to different applications and corresponding generated traffic to connected BSs. Different colors of the path correspond to different mobile applications while a

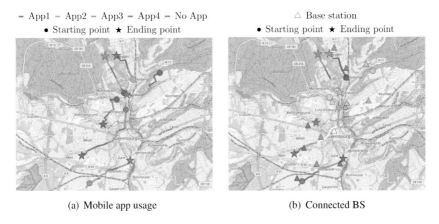

(a) Mobile app usage (b) Connected BS

Figure 1.5: Example of 5 users' trajectories with used apps and connected BSs.

citizen uses them. Red indicates that no app has been used while walking. Blue, orange, and green indicate synthetic traces and augmented reality respectively, health app, and computer intensive. Violet represents the application of video streaming characterized by real traces. Figure 1.5(b) shows with which BS (represented by a triangle) a mobile device exchanges data while walking. A mobile device automatically connects to the closest BS when it is available (for example, when it is not overloaded). In case a close BS is not available, the device connects to the second closest BS and so on.

Figure 1.6 presents the deployment of 8 EDCs with DDA and MDA approaches in Luxembourg City. Circles and stars represent BSs and EDCs respectively. BSs are assigned to an EDC of the same color. This unveils the fact that considering the mobility of citizens leads to a significantly different EDC deployment. DDA (see Fig. 1.6(a)) deploys EDCs so that all the controlled BSs are at a similar distance. The MDA approach deploys EDCs among BSs that experience higher computational demands (see Fig. 1.6(a)). Specifically, with MDA most of the EDCs tend to be deployed closer to the city center and two of them in the northeastern district of the city (Kirchberg area), which are the most important work and business districts of Luxembourg City and are very crowded during working days, especially at lunchtime.

1.20   Outage assessment

Figure 1.7(a) and Fig. 1.7(b) show the hourly outage probability in a working day for the proposed approaches with a fixed number of servers per EDC ($N_s = 10$) and a varying number too. Figure 1.7(a) illustrates the outage probability for the DDA approach. MDA clearly outperforms DDA, as shown in

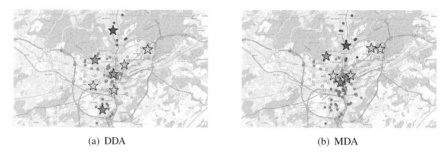

(a) DDA   (b) MDA

Figure 1.6: Distributed (DDA) and Mobility-aware (MDA) Deployment Algorithms.

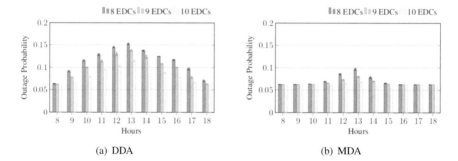

(a) DDA   (b) MDA

Figure 1.7: Outage Probability with a different number of EDCs (number of servers per EDC fixed to 10 - (a)) and with a different number of servers (number of EDCs deployed in city fixed to 8 - (b)).

Fig. 1.7(b). Interestingly, the results show that the variation of the number of EDCs shows a different behavior by the two approaches. Increasing the number of EDCs in the city makes the outage probability decrease proportionally for DDA. This is not true in MDA as having 9 or 10 EDCs makes a difference only around lunchtime, while during the day the number of EDCs does not represent a big difference. In both approaches, from hours of big workloads the outages remain similar to those due to communication delays for communications, which do not depend on the resources developed that increases the performance only in terms of computational latency.

This chapter now analyzes the impact of server allocation in the EDCs (see Subsection 1.6.4). Figure 1.8 compares DDA and MDA with the two allocation policies FNS and PNS. The number of EDCs is set to 8 and the number of servers per EDC is set to 10 for a FNS. Figure 1.8(a) compares FNS and PNS for DDA. As expected, PNS outperforms FNS for most of the time, especially during busy hours. Surprisingly, the PNS approach at 8:00 and 18:00 performs very similarly to PNS. The reason is that fewer servers are assigned to EDCs with low computa-

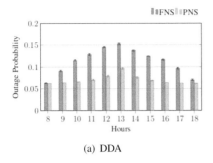 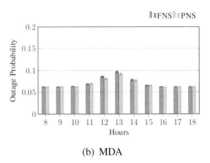

(a) DDA   (b) MDA

Figure 1.8: Total Outage Probability in a working day with FNS and PNS approach.

tional demand and it leads to an outage when other EDCs around the city do not have requests to satisfy. Thus, to reduce outages without wasting resources, one can plan a city-deployment of EDCs before deciding on how to set the number of servers per EDC. To this end, Fig. 1.8(b) shows that under the MDA approach, PNS and FNs perform very similarly at any time during the day. Interestingly, the FNS policy in MDA clearly outperforms the same policy in DDA.

1.21   Energy assessment

Figure 1.9 shows the power savings that can be achieved by powering off BSs with the EMDA approach. Specifically, a BS is powered off for a duration W when its load above the previous window's $W - 1$ is below a threshold δ of the total load of all active BSs. Then it is activated again. In these experiments, W is set to 1 hour and $\delta = 0.5\,\%$. By increasing δ and decreasing W, further energy savings can be achieved at the cost of an increase in outages. With the current setting, the higher energy-efficiency comes at no cost, for example, the outages of EMDA and MDA remain almost identical with an increase of 1.73 % on an average. Specifically, Fig. 1.9(a) shows the number of active BSs at different times of day. Clearly, at night the traffic load is lower than during day and EMDA powers offs a significant amount of BSs (down to 18 at 4 AM). Note that even during peak load (at 13 hours), the number of active BSs with EMDA is lower than with MDA. Figure 1.9(b) shows the per-hour power consumption of all the BSs. The total BSs consumption reveals how EMDA saves a significant amount of energy (on average 27 %) during the day.

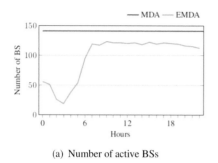

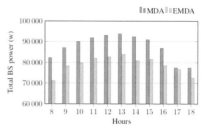

(a) Number of active BSs (b) Power consumption of all BSs per hour

Figure 1.9: Comparison between MDA and EMDA approaches.

1.22 Conclusions

This chapter tackles the problem of EDCs deployment in urban environments. By considering the citizens' mobility and their social interactions, this chapter shows that the overall performance of the MEC system can improve. This study models the computational demand and citizens mobility and formulates a problem to minimize outages, that is in the case where a user is blocked by the system because of a lack of computing resources. To solve the problem, three heuristics are proposed, one of which yields significant energy savings at the level of radio access networking with powering off BSs with low traffic and re-routing the traffic over the active BSs.

One of the contributions of this chapter is the proposal of a simulator as a MEC evaluation platform. Simulations are necessary to reach the scalability necessary to assess performance of deployments at the urban-scale level. CrowdEdgeSim is tailored to support multiple mobility models. Edge applications can be modeled parametrically or via packet traces. For the latter, this work includes video streaming with DASH and BOLA ABR method based on experiments performed on the openLEON emulator.

Acknowledgment

Dr. Fiandrino's work is supported by the Juan de la Cierva grant (FJCI-2017-32309). Dr. Guido Cantelmo's research is supported by the Marie Skłodowska-Curie grant agreement No 754462. Mr. Vitello's research is supported by the Luxembourg National Research Fund (PRIDE17/12252781/DRIVEN).

Chapter 2

Effective Data Assimilation with Machine Learning

Rossella Arcucci

2.1 Introduction

All numerical models introduce uncertainty through the selection of scales and parameters and any computational methodology contributes to uncertainty due to discretization, finite precision and accumulation of rounding-off errors. Taking these uncertainties into account is essential for any numerical simulation to be acceptable. Numerical forecasting models often use Data Assimilation methods to quantify uncertainty in the medium due to long-term analysis. Data Assimilation (DA) is the approximation of the true state of a physical system at a given time by combining time-distributed observations with a dynamic model optimally. DA incorporates observational data into a prediction model in order to improve numerically forecasted results. It allows for problems with uneven spatial and temporal data distribution and redundancy to be addressed so that models can ingest information. DA is a vital step in numerical modeling. In the past 20 years DA methodology has become the main component in the development and validation of mathematical models in meteorology, climatology, geophysics, geology and hydrology. Recently, DA has also been applied to numeri-

Data Science Institute, Imperial College London, UK.

Email: r.arcucci@imperial.ac.uk

cal simulations of geophysical applications, medicine and biological science. DA can be classically approached in two ways: as variational and filtering DA [38], [192], such as such as Kalman Filters (KF). In both cases, the methods are computed for an optimal solution: statistically, KF methods try to find a solution with minimum variance, while variational methods compute a solution that minimize a suitable cost function. In certain cases, the two approaches are identical and provide exactly the same solution [193]. While the statistical approach is often complex and time-consuming, it can provide a richer information structure. Variational approaches are relatively rapid and robust instead [38]. Recently hybrid approaches [221] have become very popular as they combine the two approaches into a single one taking advantage of the relative rapidity and robustness of variational approaches, and at the same time, obtaining an accurate solution thanks to the statistical approach.

In the past few years, both Kalman filters and variational approaches have increased in sophistication to better fit their application requirements and circumvent their implementation issues. Nevertheless, these approaches are incapable of overcoming their unrealistic assumptions fully, particularly, linearity, normality and zero error covariances.

With the rapid developments in recent years, deep learning shows great capability in approximating nonlinear systems, and extracting high-dimensional features. Machine learning algorithms are capable of assisting or replacing the traditional methods in making forecasts, without the assumptions of the conventional methods. On the other side, the training data provided for deep learning technologies, includes several numerical, approximated and rounding off errors which are entrained in the deep learning forecasting model. This means that, in some safety and security-sensitive scenarios, deep learning is still not qualified to avoid unpredictable risks. Integration of machine learning with data assimilation increases the reliability of prediction, reducing errors by including information with an actual physical meaning from observed data. The resulting cohesion of machine learning and data assimilation is then blended in a future generation of fast and more accurate predictive models. This integration is based on the idea of using machine learning to learn the past experiences of an assimilation process. This follows Bayesian approach principle and meets the needs of both DA and ML in Data Learning [213] technologies.

In DA, one makes repeated corrections to data during a single run, to bring the code output into agreement with the latest observed data. In operational forecasting there is insufficient time to restart a run from the beginning with new data hence DA should enable real-time utilisation of data to improve predictions. This mandates the choice of efficient methods to opportunely develop and implement DA models. Due to the scale of the forecasting areas and the number of state variables used to model the real world, DA is a large scale problem that should be solved in near real-time. We will show how to accelerate DA simulations introducing Machine Learning models and Domain Decomposition methods in the DA process at the top layer of the mathematical stack.

2.1.1 Related works and novelty of the presented approach

A variational DA model (VarDA) has been introduced in [27] in which it has been shown that the using an optimal space for performing DA can reduce the execution-time. However, the interface between this method and the Computational Fluid Dynamic (CFD) model presents a big bottleneck due to the accumulation of data after the CFD simulation to run DA. A parallel version of [27], based on an adaptive domain decomposition approach, has been presented in [28]. Parallelisation of data assimilation has also been previously used in oceanography. Previous works on parallelised data assimilation include domain decomposition (DD) in a regional ocean model of Australia [275], of the Caspian sea [24] and a parallel implementation of data assimilation for operational oceanography by [348, 101]. A domain decomposition approach for data assimilation has been presented in [26] where the authors implemented a geographical decomposition made of sub-domains (of the same fixed size) with overlapping.

Studies that have explored a data-driven approach for predicting climate variables climate variables include pattern searching [44], regression [247], and support vector machines [222] among others. Artificial neural networks (NN) are of particular interest since they provide ease of setting up, and can learn subtle features from the data fed into them [139]. Therefore, a mix of data-driven and data assimilation methods could be an attractive solution.

Neural networks for correction of error in forecasting have been previously studied in [41, 42, 43]. However, in this literature, the error correction by them does not have a direct relation with the updated model system in each step and the training is not based on the results of the assimilation process. In [399] the authors describe a framework for integration of NN with physical models by DA algorithms. In this approach, the NNs are iteratively trained when observed data are updated. However, this approach presents a limit due to the time complexity of the numerical models involved, which make it impossible to use the model for big data problems. An approach using a reduced order model (ROM), DA and NN has been reported in [378] and in [290] to forecast ocean variables.

Here we present an optimal approach combining DA and ML in an optimal reduced space and defined on a Domain Decomposition.

This results in a Domain Decomposition Non-intrusive Reduced Order Modelling method [30, 379] with Domain Decomposition Data Assimilation [24, 26] in a Domain Decomposition Reduced Order Data Assimilation model in order to achieve both accuracy and efficiency in forecasting. This approach improves

- the accuracy of a standard ROM model by introducing information from observed data using a DA process.

- the efficiency of the DA process in the pre-processing phase where ROM's results can be used to train the background error covariance matrices. This results in a strong reduction of the overall execution time.

These results are in accordance with the ETP4HPC Strategic Research Agenda recommendations [11]: "*the key architectural changes of exascale machines will force changes throughout the overall layers of the so called mathematics stack, including the mathematical modeling, in ways that cannot be completely hidden from the associated numerical solvers situated at the bottom layer of the math stack*". We face a problem of a big dimension introducing a reduced order model combined with DA and ML. According to the ETP4HPC requirements, the approach we proposed is scalable and, then, easily integrated with new Big Data technologies and tools.

The rest of the chapter is structured as follows: Section 2.2 introduces Data Assimilation, and Section 2.3 presents an optimal approach to reduced space. Section 2.4 describes our proposal and Section 2.5 presents and discusses the experimental results. We conclude the chapter with a summary and a brief discussion on future work.

2.2 Data assimilation

Data Assimilation (DA) is an approximation of the true state of some physical system at a given time by combining time-distributed observations with a dynamic model optimally. DA can be classically approached in two ways: as variational DA [193] or filtering. One of the best known tools for the filtering approach is the Kalman filter (KF) [192].

Let $\Omega = \{x_j\}_{j=1,\dots,N}$ be a spatial domain and let $[0, T_1] = \{t_k\}_{k=0,1,\dots,M}$ be a time window. Let

$$\mathbf{u}_k^{\mathcal{M}} \equiv \mathbf{u}(t_k) \in \mathfrak{R}^N \tag{2.1}$$

be a vector denoting the state of a dynamical system (it is often called background). At time t_k it is

$$\mathbf{u}(t_k) = \mathcal{M}(\mathbf{u}, t_{k-1}) + b_t \tag{2.2}$$

with $\mathcal{M} : \mathfrak{R}^N \mapsto \mathfrak{R}^N$, the evolutive model is often called the forecasting model.

At each time step t_k, let,

$$\mathbf{v}_k = \mathcal{H}_k(\mathbf{u}_k) + r_t \in \Re^p \tag{2.3}$$

be the vector of observations where $\mathcal{H}_k : \Re^N \mapsto \Re^p$ is a non-linear interpolation operator collecting the observations at time t_k.

The aim of the DA problem is to find an optimal tradeoff between the current estimate of the system state (the background) in (2.1) and the available observations v_k in (2.3).

Let $\mathbf{H}_k \in \Re^{nobs \times n}$ represent a linear approximation of the Jacobian of $\mathcal{H}_k$. The random vectors b_t and r_t represent the modeling and the observation errors respectively. They are assumed to be independent, white-noise processes with normal probability distributions

$$b_t \sim \mathcal{N}(0, \mathbf{B}_t), \qquad r_t \sim \mathcal{N}(0, \mathbf{R}_t) \tag{2.4}$$

where B_t and R_t are covariance matrices of the modeling and observation errors respectively.

Let the measurements error covariance matrix $\mathbf{R} \in \Re^{(m \times n_{obs}) \times (m \times n_{obs})}$, which describes the probability distribution function (pdf) of measurement errors, be defined as follows

$$\mathbf{R} = diag\,(\mathbf{R_k}), \quad \mathbf{R_k} := \sigma_0^2 \mathbf{I}, \tag{2.5}$$

with $0 \leq \sigma_0^2 \leq 1$ and $\mathbf{I} \in \Re^{n_{obs} \times n_{obs}}$ be the identity matrix.

The background error covariance matrix $\mathbf{B} \in \Re^{n \times n}$, which describes the pdf of background errors is,

$$\mathbf{B} = \mathbf{VV}^T \quad , \tag{2.6}$$

with

$$\mathbf{V} = \mathbf{X} - \bar{\mathbf{X}} \in \Re^{n \times m}, \tag{2.7}$$

where each row X_i of $\mathbf{X}$ has mean $E[X_i] = \{\bar{\mathbf{X}}_i\}_{i=1,\dots,n}$ and $\bar{\mathbf{X}} = (\bar{\mathbf{X}}_i)_{i=1,\dots,n}$.

The DA process can then be described as following:

Definition 2.1 3D Variational (3DVAR) Data Assimilation   For a fixed time $t_k = t_0$, the 3DVAR computational model is a non-linear least square problem:

$$\mathbf{u}_k^{DA} = argmin_{\mathbf{u}} J(\mathbf{u}) \tag{2.8}$$

with

$$J(\mathbf{u}) = \alpha \|\mathbf{u} - \mathbf{u}_k^{\mathcal{M}}\|_{\mathbf{B}^{-1}}^2 + \|\mathbf{H}_k \mathbf{u} - \mathbf{v}_k\|_{\mathbf{R}^{-1}}^2 \tag{2.9}$$

where, for any vector $w \in \Re^n$ and $q \in \Re^{m \times n_{obs}}$, $\|w\|_{\mathbf{B}^{-1}} = w^T \mathbf{B}^{-1} w$ and $\|w\|_{\mathbf{R}^{-1}} = w^T \mathbf{R}^{-1} w$. Parameter $\alpha > 0$ denotes the regularization parameter. In general, operational DA software assumes, $\alpha = 1$. Choosing $\alpha = 1$ can be considered as giving the same relative weight to the observations in comparison to the background state.

If eq. (2.9) is linearised around the background state [228] the VarDA problem is formulated by the following form:

$$\delta \mathbf{u}_k^{DA} = argmin_{\delta \mathbf{u}} J(\delta \mathbf{u}) \tag{2.10}$$

with

$$J(\delta \mathbf{u}) = \frac{1}{2} \alpha \delta \mathbf{u}^T \mathbf{B}^{-1} \delta \mathbf{u} + \frac{1}{2} (\mathbf{H} \delta \mathbf{u} - \mathbf{d}_k)^T \mathbf{R}^{-1} (\mathbf{H}_k \delta \mathbf{u} - \mathbf{d}_k) \tag{2.11}$$

where $\mathbf{d}_k = [\mathbf{v}_k - \mathbf{H}_k \mathbf{u}_k^{\mathcal{M}}]$ is the misfit, $\mathbf{H}$ here are the linearised observational and model operators evaluated at $\mathbf{u} = \mathbf{u}_k^{\mathcal{M}}$, and $\delta \mathbf{u} = \mathbf{u} - \mathbf{u}_k^{\mathcal{M}}$ are the increments.

In equation (2.11), the minimisation problem is defined on the field of increments [86]. In order to avoid the inversion of $\mathbf{B}$, as $\mathbf{B} = \mathbf{V}\mathbf{V}^T$ (see equation (2.6)), the minimisation can be computed with respect to a new variable $\mathbf{w} = \mathbf{V}^+ \delta \mathbf{u}$ [228], where $\mathbf{V}^+$ denotes the generalised inverse of $\mathbf{V}$, yielding to:

$$\mathbf{w}^{DA} = argmin_{\mathbf{w}} J(\mathbf{w}) \tag{2.12}$$

with

$$J(\mathbf{w}) = \frac{1}{2} \alpha \mathbf{w}^T \mathbf{w} + \frac{1}{2} (\mathbf{H}\mathbf{V}\mathbf{w} - \mathbf{d})^T \mathbf{R}^{-1} (\mathbf{H}\mathbf{V}\mathbf{w} - \mathbf{d}). \tag{2.13}$$

As the background error covariance matrix is ill-conditioned [267], in order to improve the conditioning, only Empirical Orthogonal Functions (EOFs) of the first largest eigenvalues of the error covariance matrix are considered. Since its introduction to meteorology by Edward Lorenz [229], EOFs analysis has become a fundamental tool in atmosphere, ocean, and climate science for data diagnostics and dynamical mode reduction. Each of these applications exploits the fact that EOFs allow the decomposition of a data function into a set of orthogonal functions, which are designed so that only a few of them are needed in lower-dimensional approximations [153]. Furthermore, since EOFs are the eigenvectors of the error covariance matrix [152], its condition number is reduced as well. Nevertheless, the accuracy of the solution obtained by truncating EOFs exhibits a severe sensibility to the variation of the value of the truncation parameter, so that a suitably chosen number of EOFs is strongly recommended. This issue introduces a severe drawback to the reliability of EOF truncation, hence to the usability of the operative software in different scenarios [152, 25]. To face this issue, we set the optimal choice of the truncation parameter as a trade-off between efficiency and accuracy of the DA algorithm as introduced in [27].

2.3 Reduced space

We adopt the EOFs method in order to reduce the ill-conditioning and remove the statistically less significant modes which could add noise to the estimate obtained from DA. EOFs are the eigenvectors of the error covariance matrix [152] computed by a Singular Value Decomposition:

Definition 2.2 Singular Value Decomposition Let $A \in \Re^{N \times M}$ where $N \geq M$ and let

$$A = U\Sigma W^T \tag{2.14}$$

be the singular value decomposition (SVD) of A where $U \in \Re^{N \times N}$ and $W \in \Re^{M \times M}$ are orthogonal (or orthonormal) matrices and

$$\Sigma = diag(\sigma_j)_{j=1,\ldots,N}$$

where singular values σ_j appear in decreasing order:

$$\sigma_1 \geq \sigma_2 \geq \ldots \geq \sigma_N > 0.$$

If A is a matrix of an over-determined linear system then the discrete problem is ill posed, it is needed to filter out the contribution to the solution corresponding to the smallest singular values [67, 154]. In this case, it might make sense to look at the matrix numerical rank and Singular Value Decomposition (SVD) enables us to deal with this concept. Filtering can be introduced recurring to the Truncated Singular Value Decomposition as given in the following definitions:

Definition 2.3 Truncated Singular Value Decomposition Let $A = U\Sigma W^T$ be the SVD of A as in (2.14). Let $\Phi_{trnc} \in \Re^{N \times N}$ be a matrix such that

$$\Phi_{trnc} = diag(\underbrace{1,1,1,\ldots 1}_{trnc},0\ldots,0), \tag{2.15}$$

with $1 \leq trnc \leq N$. Then the matrix

$$A^{trnc} := U\Phi_{trnc}\Sigma W^T, \tag{2.16}$$

is the truncated SVD (TSVD) matrix for S.

In order to improve the conditioning, only the first largest eigenvalues of the error covariance matrix are considered. Even if the employment of EOFs which strongly reduce the dimension, alleviate the computational cost, nevertheless, a consequence is that important information is lost [61]. This issue introduces a severe drawback to the reliability of the EOFs if the truncation parameter is not properly chosen.

As it is known that the numerical error which propagates into the DA solution is influenced by the condition number [25], a proper value of the truncation parameter should *minimise the condition number*. However, to be sure that the preconditioned problem does not depart too much from the original problem, the optimal truncation parameter should also *minimise a Relative Preconditioning Error* (RPE) which provides an estimate of how much the preconditioned problem differs from the starting problem as defined in Definition 2.4.

Definition 2.4 Relative Preconditioning Error (RPE)   Let E_τ be the relative Preconditioning Error which provides an estimate of how much the preconditioned problem differs from the starting problem and defined as:

$$E_\tau = \frac{\|\Sigma - \Phi_\tau\|_\infty}{\|\Sigma\|_\infty}. \tag{2.17}$$

The proper choice of the truncation parameter is usually related to a reference exact solution [67]. However, in operational Data Assimilation, the knowledge of this reference solution represents a strong constraint. To avoid this drawback, we assume an estimation of the truncation parameter which is independent of the knowledge of an exact solution [27]. In fact, an optimal truncation parameter σ_{opt} should be picked to minimise both:

- the condition number of $\mathbf{V}$ after the preconditioning [154]

$$\mu(\mathbf{V}_\tau) \simeq \frac{\sigma_1}{\sigma_\tau}, \tag{2.18}$$

- the Relative Preconditioning Error (2.17);

The asymptotic behaviour of the condition number in (2.18) is

$$\lim_{\sigma_\tau \to 0} \mu(\mathbf{V}_\tau) = +\infty, \qquad \lim_{\sigma_\tau \to +\infty} \mu(\mathbf{V}_\tau) = 0, \tag{2.19}$$

and of the Relative Preconditioning Error is

$$\lim_{\sigma_\tau \to 0} E_\tau = \frac{\|\Sigma - \mathbf{I}\|_\infty}{\|\Sigma\|_\infty} \simeq 1 - \frac{1}{\sigma_1}, \qquad \lim_{\sigma_\tau \to +\infty} E_\tau = \frac{\|\Sigma\|_\infty}{\|\Sigma\|_\infty} = 1 \tag{2.20}$$

As σ_τ is subject to the constraints $\sigma_1 \geq \sigma_\tau \geq \sigma_N$ [67], from equation (2.19), we find that the smallest value of the condition number is obtained for $\sigma_\tau \simeq \sigma_1$. From equation (2.20), however, the smallest error is obtained for $\sigma_\tau \simeq \sigma_N$.

Due to this difference in the asymptotic behaviour of the two functions E_τ and $\mu(\mathbf{V}_\tau)$, the optimal value is the minimum of the function

$$f(\sigma_\tau) = E_\tau + \mu(\mathbf{V}_\tau) \tag{2.21}$$

which is [27]:

$$\sigma_{opt} = argmin \; f(\sigma_\tau) = \sqrt{\sigma_1} \tag{2.22}$$

The equation (2.22) is the condition we assume to select the optimal truncation parameter as implemented in Algorithm 4.

Algorithm 4 Optimal Reduced Space algorithm

1: Input: $\{u_{ki}^M\}_{k=0,\ldots,M \; i=1,\ldots,N}$ ▷ temporal sequence of historical data
2: Compute $\{\bar{u}_i\}_{i=1,\ldots,N} = mean_{k=0,\ldots,M}\left(\{u_{ki}^M\}_{i=1,\ldots,N}\right)$ ▷ compute the mean over the temporal sequence
3: Compute $\mathbf{V} = \{V_{ki}\}_{k=0,\ldots,M \; i=1,\ldots,N}$ such that $V_{ki} = u_{ki}^M - \bar{u}_i$ ▷ compute the matrix $\mathbf{V}$
4: Compute $\mathbf{V} = \mathbf{U\Sigma W}^T$.   ▷ compute the EOFs of the matrix $\mathbf{V}$
5: Set $e = \sqrt{\sigma_1}$
6: Set $\tau = 1$
7: for $i = 2,\ldots,M$ ▷ compute the truncation parameter
8:     if $(\sigma_\tau < e)$ then
9:        $\tau = \tau + 1$
10:     end if
11: end for
12: Compute $\mathbf{V}_\tau = EOFs(\mathbf{V}, \tau)$. ▷ Truncated SVD regularised matrix in (2.16)

2.4 Effective data assimilation with machine learning

In this section a synthetic mathematical formalization of the DA model based on a Domain Decomposition approach and using a Machine Learning method is presented. Local functions are introduced. These functions are defined on subdomains which constitute a partitioning of the domain with overlapping. First, some preliminary definitions are introduced in order to help the mathematical description of the restriction and the extension of the function among the subdomains and the global domain respectively.

Definition 2.5 Domain Decomposition   The set of overlapping sub-domains

$$DD(\Omega) = \{\Omega_i\}_{i=1,\ldots,p} \tag{2.23}$$

is a decomposition of the domain $\Omega \subset \Re^N$, if $\Omega_i \subset \Re^{r_i}$, $r_i \leq N$ for $i = 1, \ldots, p$, it is such that

$$\Omega = \bigcup_{i=1}^{p} \Omega_i \qquad (2.24)$$

with

$$\Omega_i \cap \Omega_j = \Omega_{ij} \neq \emptyset$$

♠

Let us define the Restriction Operator RO_i and Extension Operators EO_i. Let $t \in [0, T]$, associated with decomposition (2.24) and we define the Restriction Operator for a function belonging to the Hilbert space $\mathcal{K}(\Omega)$:

Definition 2.6 Restriction Operator *Let us define:*

$$RO_i : \mathcal{K}(\Omega) \mapsto \mathcal{K}(\Omega_i)$$

such that:

$$RO_i(\mathbf{u}(x)) \equiv \mathbf{u}(x), \qquad x \in \Omega_i$$

Moreover, for simplicity of notations, we let:

$$\mathbf{u}^{RO_i}(x) \equiv RO_i[\mathbf{u}(x)]$$

♠

In this respect we define the Extension Operator also given a set of p functions $\mathbf{u}_i$, $i = 1, \ldots, p$, each belonging to the Hilbert space $\mathcal{K}([0, T] \times \Omega_i)$:

Definition 2.7 Extension Operator *Let us define:*

$$EO_i : \mathcal{K}(\Omega_i) \mapsto \mathcal{K}(\Omega)$$

such that:

$$EO_i(\mathbf{u}_i(x)) = \begin{cases} \mathbf{u}_i(x) & x \in \Omega_i \\ 0 & elsewhere \end{cases}$$

Moreover, for simplicity of notations, we let:

$$\mathbf{u}_i^{EO}(x) \equiv EO_i[\mathbf{u}_i(x)]$$

♠

We observe that, for any function $\mathbf{u} \in \mathcal{K}(\Omega)$, associated to the decomposition (2.24), it holds that

$$\mathbf{u}(x) = \sum_{i=1,p} EO_i \left[\mathbf{u}_i^{RO}(x) \right]. \tag{2.25}$$

Given p functions $\mathbf{u}_i(x) \in \mathcal{K}(\Omega_i)$, their summation

$$\sum_{i=1,p} \mathbf{u}_i^{EO}(x) \tag{2.26}$$

defines a function $\mathbf{u} \in \mathcal{K}(\Omega)$ such that:

$$RO_j[\mathbf{u}(x)] = RO_j \left[\sum_{i=1,p} \mathbf{u}_i^{EO}(x) \right] = \mathbf{u}_j(x). \tag{2.27}$$

Remark 2.1 We will use the same notation to denote the restriction/extension operators acting on points and vectors. If

$$\Omega = \{x_1, x_2, \ldots, \ldots, x_N\} \subseteq \Re^{N \times N}$$

and

$$\mathbf{u} = (\mathbf{u}(x_1), \mathbf{u}(x_2), \ldots, \mathbf{u}(x_N)) \in \Re^N,$$

where $\mathbf{u} : \Re^N \mapsto \Re$, let us assume that Ω can be decomposed into a sequence of $p \geq 1$ overlapping sub-domains Ω_i such that

$$\Omega = \bigcup_{i=1}^{p} \Omega_i$$

where $\Omega_i \subset \Re^{r_i \times N}$ and $r_i \leq N$. Hence

$$RO_i(\mathbf{u}) \equiv \mathbf{u}^{RO_i} \equiv (\mathbf{u}(x_i))_{x_i \in \Omega_i}, \quad \mathbf{u}^{RO_i} \in \Re^{r_i}.$$

In this respect we define the adjoint Extension Operator also. If $\mathbf{u} = (\mathbf{u}(z_i))_{z_i \in \Omega_i}$, it is

$$EO_i(\mathbf{u}) = \begin{cases} \mathbf{u}(z_k) & z_k \in \Omega_i \\ 0 & elsewhere \end{cases}$$

and $EO_i(z) \equiv \mathbf{u}^{EO_i} \in \Re^N$. ∎

Let $J(\mathbf{u}, \mathbf{u}^{\mathcal{M}}, \mathbf{v}, \mathbf{B}, \mathbf{R}, \Omega)$ denote the DA functional as defined in (2.9). We can extend the concepts of Extension and Restriction to the functional.

Definition 2.8 Functional restriction *We generalize the definition of the restriction operator RO_i acting on J, as follows:*

$$RO_i : J(\mathbf{u}, \mathbf{u}^{\mathcal{M}}, \mathbf{v}, \mathbf{B}, \mathbf{R}, \Omega) \mapsto J(\mathbf{u}^{RO_i}, RO_i(\mathbf{u}^{\mathcal{M}}), RO_i(\mathbf{v}), RO_i(\mathbf{B}), RO_i(\mathbf{R}), \Omega_i)$$
$$(2.28)$$

♠

For simplicity of notations we let $u_i = u^{RO_i}$ and:

$$J(\mathbf{u}^{RO_i}, RO_i(\mathbf{u}^{\mathcal{M}}), RO_i(\mathbf{v}), RO_i(\mathbf{B}), RO_i(\mathbf{R}), \Omega_i) \equiv J_{\Omega_i}$$

Definition 2.9 Functional extension We generalize the definition of the extension operator EO_i acting on J_{Ω_i} as

$$EO_i : J(\mathbf{u}^{RO_i}, RO_i(\mathbf{u}^{\mathcal{M}}), RO_i(\mathbf{v}), RO_i(\mathbf{B}), RO_i(\mathbf{R}), \Omega_i) \mapsto J(\mathbf{u}, \mathbf{u}^{\mathcal{M}}, \mathbf{v}, \mathbf{B}, \mathbf{R}, \Omega)$$

where

$$EO_i[J_{\Omega_i}] : \mathbf{w} \in \Re^N \mapsto \begin{cases} J(EO_i(\mathbf{w}^{RO_k})) & k \in \Omega_i \\ 0 & k \notin \Omega_i \end{cases}$$
$$(2.29)$$

♠

We observe that
$$J(\mathbf{u}, \mathbf{u}^{\mathcal{M}}, \mathbf{v}, \mathbf{B}, \mathbf{R}, \Omega) = \sum_{i=1,p} J_{\Omega_i}^{EO_i} .$$
$$(2.30)$$

We are now able to define the *local* function. A local function describes the local problem on a sub-domain Ω_i of the domain decomposition. It is obtained by applying the restriction operator to the function J and by adding a *local* constraint to it. This is in order to enforce the continuity of each solution of the local DA problem onto the overlap region between adjacent domains Ω_i and Ω_j.

Definition 2.10 Local function The operator J_i defined as:

$$J(\mathbf{w}_i) = \frac{1}{2}\alpha\mathbf{w}_i^T\mathbf{w}_i + \frac{1}{2}(\mathbf{H}_i\mathbf{V}_i\mathbf{w}_i - \mathbf{d}_i)^T\mathbf{R}_i^{-1}(\mathbf{H}_i\mathbf{V}_i\mathbf{w}_i - \mathbf{d}_i).$$
$$+\mu\,(\mathbf{w}_i/\Omega_{ij} - \mathbf{w}_j/\Omega_{ij})^T\mathbf{V}_{ij}(\mathbf{w}_i/\Omega_{ij} - \mathbf{w}_j/\Omega_{ij})$$
$$(2.31)$$

where $\mathbf{w}_i$, $\mathbf{V}_i$, $\mathbf{R}_i$ and $\mathbf{d}_i$ are restrictions on Ω_i of vectors and matrices in (2.13), and $\mathbf{w}_i/\Omega_{ij}$, $\mathbf{w}_j/\Omega_{ij}$, $\mathbf{V}_{ij}$ are restrictions on $\Omega_{ij} = \Omega_i \cap \Omega_j$ the quantities defined in (2.13) and is named the local funtion.

♠

Parameter μ is equal to 1: it ensures the continuity of each solution of the local problem onto the overlap region between adjacent domains.

In this section the accuracy of the introduced model is also studied. An important point is to ensure that the introduction of a decomposition in the dataset does not introduce errors which affect the accuracy of the solution. A theorem which ensures that the solution obtained by the parallel algorithm is the same as obtained by the sequential one is proved here.

We observe that, as is for J, each J_i is a quadratic but not necessarily a convex one. This shows that the unicity of the minimum is not satisfied. Let

$$\mathbf{w}_i^m = argmin_{\mathbf{w}_i} J_i(\mathbf{w}_i) \tag{2.32}$$

be the global minimum of J_i, $\forall i : \Omega_i \in DD(\Omega)$, and the following results hold

Theorem 2.1

If $DD(\Omega)$ is a decomposition of Ω as in Definition 2.5, then, let

$$\mathbf{w}^m = \sum_{i=1,\dots,p} (\mathbf{w}_i^m)^{EO_i}, \tag{2.33}$$

where $\mathbf{w}_i^m$ is defined in (2.32), it follows that:

$$\mathbf{w}^m = \mathbf{w}^{DA},$$

with $\mathbf{w}^{DA}$ defined in (2.12).

Proof: Let $\mathbf{w}_i^m$ be the minimum of J_i on Ω_i as defined in (2.32), it is

$$\nabla J_i[\mathbf{w}_i^m] = 0, \quad \forall i : \Omega_i \in DD(\Omega). \tag{2.34}$$

From (2.34) it follows that $\sum_i \nabla J_i[\mathbf{w}_i^m] = 0$ which give from the property of the gradient gives:

$$\nabla \sum_i J_i[\mathbf{w}_i^m] = 0 \tag{2.35}$$

Let EO_i be the extension function as defined in (2.29), and from (2.35) it follows that,

$$\nabla \sum_i EO_i\,(J_i[\mathbf{w}_i^m]) = 0 \tag{2.36}$$

which from the (2.30) gives:

$$\nabla J[\sum_i (\mathbf{w}_i^m)] = 0 \tag{2.37}$$

then, from (2.37) follows that $\sum_i(\mathbf{w}_i^{m^{EO_i}})$ is a stationary point for J. As $\mathbf{w}^{DA}$ as defined in (2.12) is the global minimum, it follows that:

$$J(\mathbf{w}^{DA}) \leq J\left(\sum_i(\mathbf{w}_i^{m^{EO_i}})\right), \qquad \forall i = 1, \dots, p \tag{2.38}$$

then

$$J(\mathbf{w}^{DA}) \leq J\left(argmin_{\mathbf{w}}J(\{\sum_i(\mathbf{w}_i^{m^{EO_i}})\})\right) = J(\mathbf{w}^m) \tag{2.39}$$

so the condition $J(\mathbf{w}^{DA}) \leq J(\mathbf{w}^m)$ is satisfied. We proved that $\mathbf{w}^{DA} = \mathbf{w}^m$, then $J(\mathbf{w}^{DA}) = J(\mathbf{w}^m)$, by reduction to the absurd. Infact, by assuming

$$J(\mathbf{w}^m) < J(\mathbf{w}^{DA}), \tag{2.40}$$

in Ω the (2.40) gives $J(\mathbf{w}^{DA}(x_j)) < J(\mathbf{w}^m(x_j))$ for all $x_j \in \Omega$. In particular, let Ω_i be a subset of Ω, then $J(\mathbf{w}^{DA}(x_j)) < J(\mathbf{w}^m(x_j))$, for all $x_j \in \Omega_i$. Hence, from the definition of RO_i in (2.28) and by assuming the (2.40) we have

$$J_{\Omega_i}(RO_i(\mathbf{w}^{DA})) < J_{\Omega_i}(RO_i(\mathbf{w}^m)), \qquad \forall i : \Omega_i \subset \Omega \tag{2.41}$$

which from (2.33) gives:

$$J_{\Omega_i}(RO_i(\mathbf{w}^{DA})) < J_{\Omega_i}(\mathbf{w}_i^m). \tag{2.42}$$

Equation (2.42) is absurd. In fact, if $RO_i(\mathbf{w}^{DA}) \neq \mathbf{w}_i^m$, the (2.42) says that there exists a point such that the value of the function at that point is smaller than the value of the function at the point of the global minimum. Thus the theorem is proved.

This result ensures that $\mathbf{w}^{DA}$ (the global minimum of J) can be obtained by patching together all the vectors $\mathbf{w}_i^m$ (global minimums of the operators J_{Ω_i}), that is, by using the domain decomposition, the global minimum of the operator J can be obtained by patching together the minimums of the *local* functionals J_{Ω_i}. This result has important implications from the computational viewpoint as it will be explained in the next section.

After the decomposition of the domain as defined in (2.23), each subdomain Ω_i, is optimally reduced by referring to the TSVD as defined in (2.16) and put into effect in Algorithm 4. In the reduced space, the data is trained to estimate a forecasting function G defined in the reduced space. A recurrent neural network (RNN) can be used to train, and therefore predict, the temporal dependency of the background state of the model. We introduce a neural network G_ω with a training target:

$$\mathbf{u}_k = G_\omega\left(\mathbf{u}_{k-1}\right). \tag{2.43}$$

We train G_ω which is described in the iterative form $G_{\omega,i} \circ G_{\omega,i-1} \circ \ldots \circ G_{\omega,1}$ where $G_{\omega,i}$ is defined as the network trained in ith iteration and $\circ$ denotes the composition function. Let G_i be the model implemented in the ith iteration, it is:

$$G_i = G_{\omega,i} \circ G_{\omega,i-1} \circ \ldots \circ G_{\omega,1}, \tag{2.44}$$

from which we get, $G_i = G_{\omega,i} \circ G_{i-1}$, and $G = G_M$.

For iteration $i > 1$, the training set for $G_{\omega,i}$ is $\{(\mathbf{u}_k, \mathbf{u}_{k+1})\}$, which capitalises on the $\{(\mathbf{u}_{k-1}, \mathbf{u}_k)\}$ from the last updated model G_i. The RNN is specifically indicated as [187]

$$\begin{aligned} \mathbf{h}_t &= tanh(w_1[\mathbf{u}_k, \mathbf{h}_{t-1}] + b_1) \\ \mathbf{u}_k &= w_2(tanh(w_1[\mathbf{u}_{k-1}, \mathbf{h}_{t-1}] + b_1)) + b_2 \end{aligned} \tag{2.45}$$

where $\mathbf{h}_t$ denotes the hidden layers [109] for $t = 1, \ldots, M$ and b_1, b_2 and w_i, w_2 denote the coefficients and the weights of the network respectively.

In next Section, we validate the introduced model by experimental results provided for pollutant dispersion within an urban environment [29, 32].

2.5 Testing and results

The capability of our model has been estimated using an urban environment located in London South Bank University (LSBU) area (London, UK) shown in Figure 2.1. The computational domain has a size of $[0, 2041] \times [0, 2288] \times [0, 250]$

Algorithm 5 Effective DA

Input: v_i and u_i^M Define H_i ▷ interpolation operator
Compute $d_i \leftarrow v_i - H_i u_i^M$ ▷ compute the misfit
Define R_i ▷ covariance matrix of the observed data v_i
Define V_i ▷ deviance matrix of background data
Define the initial value of δu_i^{E-DA}
Compute V_i ▷ using Algorithm 4
  Compute $w_i \leftarrow V_i^T \delta u_i^{E-DA}$
While(Convergence on w_i is obtained)
Compute $J_i \leftarrow J_i(w_i)$
Compute $gradJ_i \leftarrow \nabla J_i(w_i)$
Compute new values for w_i ▷ L-BFGS step
Compute $u_i^{E-DA} \leftarrow u_i^M + V_i w_i$
return: u^{E-DA}

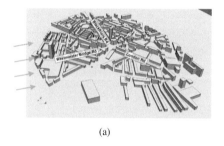 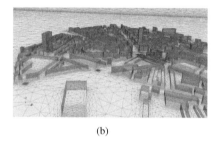

(a)   (b)

Figure 2.1: (a) London South Bank University (LSBU) test case area. The red sphere denotes the location of the pollution source and the blue arrows denote the wind direction. (b) surface mesh of the test site.

(metres). This work uses the 3D non-hydrostatic Navier-Stokes equations as the full physical system,

$$\nabla \cdot \mathbf{u} = 0, \tag{2.46}$$

$$\frac{\partial \mathbf{u}}{\partial t} + \mathbf{u} \cdot \nabla \mathbf{u} = -\nabla p + \nabla \cdot \tau, \tag{2.47}$$

where $\mathbf{u} \equiv (u, v, w)^T$ is the velocity, $p = \tilde{p}/\rho_0$ is the normalised pressure ($\tilde{p}$ being the pressure and ρ_0 the constant reference density) and τ denotes the stress tensor.

Simulations were carried out for the study area using Fluidity, an open-source, finite-element, fluid dynamics model [17]. The dispersion of pollutant is described by the classic advection-diffusion equation with the pollutant concentration treated as a passive scalar. A source term was added to the advection-diffusion equation to mimic a constant release of pollutant generated by traffic at a busy intersection for example. The location of the point source is depicted by the red sphere in the Figure 2.1(a). The time step was adaptive based on the Courant (CFL) number defined by the user, and the Crank-Nicholson scheme was used for the time discretization [31, 126]. The mesh is shown in Figure 2.1(b). The outlet boundary condition was defined by a zero-pressure (no-stress) condition; perfect slip boundary conditions were applied at the top and on the sides of the domain and no-slip boundary conditions were applied on all building facades and the bottom surface of the domain. A synthetic incoming-eddy method was used at the inlet to resemble the behaviour of the boundary layer. The mean velocity profile was presented as in equation (2.48):

$$(u, v, w) = \left(0.97561 \ln \left(\frac{z}{0.01} \right), 0, 0 \right) \tag{2.48}$$

where z denotes the height (in m). The inlet length-scale L and Reynolds stresses Re are kept constant and equal to 100 m and 0.8 respectively, for the diagonal components, and zero elsewhere. Zero velocity is prescribed at the bottom and on the wall boundaries. A zero stress condition is set with $p = 0$ at the outlet boundary and a perfect slip condition is specified for the vertical lateral boundaries. Experiments have been conducted and tested on 3 high performance nodes equipped with bi-Xeon E5-2650 v3 CPU and 250GB of RAM with Python 3.5.

The accuracy of our model's results is evaluated by the mean squared error on each sub-domain:

$$MSE(u_j) = \frac{\|\mathbf{u}_j - \mathbf{u}_j^C\|_{L^2}}{\|\mathbf{u}_j^C\|_{L^2}} \tag{2.49}$$

computed with respect to a control variable $\mathbf{u}_j^C$ provided by the observed data, for $j = 1, \ldots, s$ and s denotes the number of sub-domains. Table 2.1 shows the values of $MSE(u^G)$ and $MSE(u^{E-DA})$ for a decomposition to, $s = 16$ sub-domains running on $p = 16$ processors. We can observe that the error decreases for each sub-domain. We observe a bigger gain in terms of accuracy reduction in sub-domains where the pollution concentration is more diffused. Fox example, the sub-domain number 11 presents a bigger gain as shown in Table 2.1, this sub-domain is the central sub-domain in the decomposition as shown in Figure 2.2 (orange colour) which represents the iso-surface of the pollutant concentration for $5.10^{-1} kg/m^3$ computed in parallel with $p = 16$ processors and generated by a point source.

We evaluated the execution time needed to compute the solution of the E-DA model using Algorithm 5. Let T_s denote the execution time of Algorithm 5 for a domain decomposition made of s sub-domains. We assume that $p = s$, where p denotes the number of processors and we suggest:

$$T_s = max\{T_{s_i}\}_{i=1,\ldots,s} \tag{2.50}$$

Table 2.1: Values of $MSE(\mathbf{u}^G)$ and $MSE(\mathbf{u}^{E-DA})$ for a decomposition made of $s = 16$ subdomains running on $p = 16$ processors.

s_i	$MSE(\mathbf{u}^G)$	$MSE(\mathbf{u}^{E-DA})$
1	1.05	0.67
2	1.51	0.98
3	0.15	0.13
4	3.18	1.55
5	1.30	0.97
6	1.43	0.74
7	0.53	0.26
8	2.62	1.30
9	1.75	1.12
10	2.80	1.47
11	2.30	0.39
12	1.93	0.91
13	1.74	1.01
14	2.32	1.24
15	1.16	0.67
16	3.032	1.14

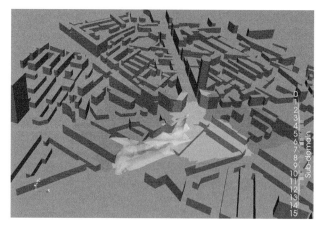

Figure 2.2: u^G: predicted pollutant iso-surface by G.

where T_{s_i} denotes the execution time for each processor in each sub-domain. The total execution time is shown in Table 2.2. There is a clear decreasing trend in the total execution time with an increasing number of processors. Table 2.2 shows the values of execution time of G and E-DA with $p = 4, 16, 32$ processors for a decomposition of $s = 4, 16, 32$ sub-domains. Table 2.3 shows the gain in terms of execution time provided by using G instead of M. As shown in this table,

Table 2.2: Values of execution times of G and E-DA for a number of sub-domains $s = 4, 16, 32$ running on $p = 4, 16, 32$ processors (plot in linear scale).

s_i	T^G (secs)	T^{E-DA} (secs)
4	2553.88	2836.98
16	1677.31	1749.89
32	977.82	1003.55

Table 2.3: Values of T^M and T^{E-DA} for a decomposition made of $s = 16$ sub-domains running on $p = 16$ processors.

s_i	T^M (secs)	T^{E-DA} (secs)
1	3327.31	1757.31
2	3359.74	1789.74
3	3363.64	1793.64
4	3390.81	1820.81
5	3394.10	1824.10
6	3396.94	1826.94
7	3399.11	1829.11
8	3401.46	1831.46
9	3409.85	1839.85
10	3416.12	1846.12
11	3423.66	1853.66
12	3445.83	1875.83
13	3450.93	1880.93
14	3453.89	1883.89
15	3465.70	1895.70
16	3480.66	1910.66

the use of G instead of M strongly impacts the efficiency of pre-processing the Data Assimilation phase for computing the covariance matrices V_j for each time step n.

2.6 Conclusions

This Chapter presented an Effective Data Assimilation with a Machine Learning model based on a Domain Decomposition approach. We proved that our approach improves both accuracy of the the ML model and efficiency of the DA process. The accuracy of the ML model is improved by introducing information from observed data exploiting the 3D variational DA process. The efficiency of the DA process is mainly improved in the pre-processing phase. In fact, we used the ML results to condition our background error covariance matrices and we have shown that this implies a strong reduction of the overall execution time. The efficiency and the accuracy of our model were discussed and tested using a 3D case of air flows and pollution transport in an urban environment. These results are in accordance with the ETP4HPC Strategic Research Agenda recommendations [11] which force changes throughout the overall layers of the so called mathematics stack, including the mathematical modeling. According to the requisite requirements, the approach we proposed is scalable and, then, easily integrated with new Big Data technologies and tools. The algorithms and the methods proposed are generic enough and can be easily used for several physical problems.

Chapter 3

Semantic Data Model for Energy Efficient Integration of Data Centres in Energy Grids

Ionut Anghel, * *Tudor Cioara, Claudia Pop* and *Marcel Antal*

3.1 Introduction

Modern Data Centres (DCs) are large-scale distributed systems with high energy consumption, which may operate in complex smart urban environments. Cloud and big data driven architectures and services have recorded a massive development in the last decade, finding a presence in both public and private sectors. However, the continuous demand of such applications leads to massive hardware infrastructure investments in DCs, a fact that has a negative impact from both economic and environmental perspectives. Studies show that DCs are responsible for 1.4% of the world energy demand and 2% of the global carbon emissions [177] that lead to global warming.

As the complexity of applications increases and their management becomes a serious issue, new features of energy situational awareness and data driven optimization are needed, so that DCs become aware of their external and internal

Department of Computer Science, Technical University of Cluj-Napoca.
* Corresponding author: ionut.anghel@cs.utcluj.ro

energy contexts and adapt their operation towards increased energy efficiency. In these circumstances, it is economically and environmentally mandatory to consider energy usage as useful contextual information and to develop energy and data driven applications capable of making optimization decisions dynamically to increase the DC energy efficiency. Their development poses research challenges regarding monitoring and representing relevant context-enriched data on one hand and using big data analytics techniques for assessing the DC energy efficiency operation, identifying energy leakages, and taking optimization actions considering various criteria on the other hand. Moreover, the perception of DCs as energy consumers, which can provide their surplus of green energy or dissipated energy to district buildings, implies additional implementation challenges.

To deal with such aspects, semantically enhanced data models need to be defined for representing data related to the energy efficiency of DC operations, enabling the usage of big data analytics and semantic reasoning techniques to make inferences from complex information and to increase the energy consumption awareness. Thus, in the context of GEYSER project [23] we had defined such a data model, implemented as an ontology, for DCs located in urban environments, as a central component that unifies the common vocabulary minimizing the energy consumption of their operations. The data model defines ontology concepts and relations for describing the internal and external energy contexts of DC operations. The DC internal context refers to all the data that characterizes the energy efficiency DC operation. The DC's external context captures data relevant for assessing the efficiency of its integration within the smart energy grid in relation to optimal shifting of energy flexibility to meet grid level defined objectives. The model is used for, semantic annotation of monitored data and to empower the data driven optimization actions for decision making through ontology-based reasoning and queries. The DC is modelled as an energy player at the crossroads of various utility lines (data, electrical, thermal, etc.), to address the issue of coordinated optimal operation of the cross-energy carrier resources such as DCs for decreasing their carbon footprint. The model described in this chapter aims to provide a new level of semantic integration for achieving the desired level of coordination.

3.2   Related work

In the research literature few approaches can be found for modelling the energy efficiency of DCs, with most of them being focused on operational efficiency of specific DC components and not providing a holistic view of the DC as an energy load resource integrated into the smart grid. They are mostly interested in defining and evaluating Key Performance Indicators (KPIs) expressing the energy efficiency of isolated DC components, failing to capture the semantics behind the monitored data and dependencies within the energy grid. A selected subset of such approaches is presented below.

In [345], the problem of correlating the cooling system operation with the work-load execution is addressed by proposing a thermal flow model allowing the adjustment of cooling intensity to minimize the DC energy consumption. On top of the model an optimization problem is presented, defined and solved that tries to minimize the cooling energy by distributing load among servers. In [366], [216] and [382] the authors investigate the integration of energy storage in such a thermal model with a view to increase the overall thermal and electrical flexibility potential. Two types of energy storage elements are modelled: batteries and thermal storage tanks for non-electrical cooling systems. For each element, the charging and recharging operations are defined, along with energy loss parameters and operational costs. Furthermore, different policies and algorithms for shaving power peaks and reducing overall DC energy costs are presented. The fluid dynamics thermal models are defined and used to study the efficient integration of thermal storage components [366] to gain flexibility. Finally, the paper presents and evaluates a smart framework that integrates and dynamically coordinates three storage elements in a DC such as thermal tanks, thermal masses and batteries. Models of the different cooling system configurations such as Computer Room Air Conditioning, free air cooling and liquid cooling are presented in [216]. A mathematical model of a DC integrating these systems is given, and an energy optimization problem is defined using fluid dynamics. The capabilities of a DC equipped with thermal storage systems and electrical energy storage systems are further extended in [382]. TEShave, a system that can shave energy peaks in order to reduce overall operational costs by intelligently integrating and dynamically coordinating storage elements is described and tested in a simulated environment. In [389] a model for a DC federation is defined. The DC is represented by a front-end proxy server that receives the tasks and a set of homogeneous backend servers that process them. The tasks are sent to the back-end cluster and placed in an execution buffer that is modelled as a queue. A delay-tolerant workload model and a cost model are presented. Three optimization decisions are analysed to estimate and reduce power costs: alternative task scheduling to the back-end servers, adjusting the number of up and running servers to accommodate them and different service level indicators to be provided for the customers. A cost-based model is used to manage the temporal and space variation of the workload task requests.

Modern DCs are integrated with utility networks both at the information and energy levels. The uptake and deployment of distributed renewable energy sources (RES) changed the energy mix increasing the proportion of green energy. For maximizing the usage of renewable energy in the local energy grid, in [242] the authors propose a power model of a DC that is described at server and network levels, and an algorithm to migrate workload among various nodes of the network. Also, several heuristics are presented as solutions to the renewable energy consumption optimization problem. In order to incorporate renewable energy into the grid it is critical to have demand-response tools. DCs are large electricity consumers, thus being a promising demand-response resource. In [227], the authors make two major contributions: they show the demand-response potential of DCs and they present a novel design for price prediction for DCs. The DC is presented as a power model, having two major components: real time workload energy consumption and delay-tolerant workload energy consumption. Furthermore, the integration of the DC with local energy generation sources is presented. Finally, the distribution network and power price models are described. Several experiments are conducted to determine the quality of price prediction tools in order to minimize operational costs for DCs. A DC model that considers local renewable energy generation is presented in [16]. The authors model a DC as a system of waiting queues for jobs, which can be real-time or delay tolerant. Furthermore, the renewable energy sources mathematical model for photovoltaic and wind sources is presented. While previous works focused on methods that suggested immediate usage of renewable energy when available, this paper proposes a method of predicting renewable energy production in order to develop an adaptive energy-aware job scheduler. In [362] the authors aim to reduce the DC cost with energy using Lyapunov optimization to optimally schedule the delay-tolerable workload to take advantage of better energy price and using batteries as a source of additional flexibility. Optimization decisions are taken regarding the amount of energy which should be drawn from the grid, the energy amount to be stored and drawn from the batteries to minimize the operational costs. A complex model of the DC operation is described in [3]. The authors model the DC as a set of heterogeneous servers, with a complementary renewable power source and two types of workloads: real-time and delay tolerant. Furthermore, they model the temporal operation of the DC, by introducing an online algorithm that is based on predictions. Finally, they compare their online budgeting algorithm with classical offline algorithms to prove the advantages of such an approach. Defining and representing data and information in a computer interpretable manner is a prerequisite for optimization algorithms to process and reason, to take decisions and generate useful feedback. Ontologies are good candidates for defining vocabularies and domain taxonomies in multiple domains while being the core component for developing complex reasoning and learning techniques and models [77] and for ultimately defining context models that can help design context aware systems [342]. [95] presents a DC ontological taxonomy defining the main concepts, their properties and their relationships. An instantiation of the meta-model can represent a DC state. The model is used for analyzing processes of multiple situations in the DC. However, no energy related information is represented in the proposed meta-model which is constructed off line using DC knowledge and updated continuously in a semi-automated manner. In [89], the authors propose a framework for low carbon management of energy resources in the context of smart grid technologies featuring an ontology for their semantic representation as Resource Description Framework (RDF) graph models. The ontology contains information regarding the energy consumption and CO_2 emissions of such grid energy resources. Energy is classified as green or brown depending on its source. The proposed model is used for the resource discovery process in DC federations, based on semantics similarity and Bayesian networks, and is primarily used for direct end-user searches, giving the most suitable green resource as per the user request. However, it does not consider additional information useful for deriving energy efficient schedules, such as the energy price, resource power consumption curves, the option of participating in an energy marketplace among others. In [148] a platform which can be used to automate DC and cloud operations is presented. It uses a data model to map various information sources (storage, computational resources, applications, to mention a few) for obtaining a snapshot of the DC. The model does not contain any energy related concepts and is mainly used for presenting information to the DC operator, and queries for information and analysis of historical data. Analysing the state-of-the-art approaches described above, we can conclude that they only partially address the DC energy data modelling, with none providing a uniform and comprehensive vision on DC energy and performance representation. Also, none of the studied models aim at capturing the efficiency of the DC integration in the smart energy grid. As opposed to these models, the proposed data model is not focusing on evaluating some predefined KPIs, but rather on capturing the semantic dependencies and relations in DC energy data and among components aiming to enact the situational awareness of their energy efficiency in operation.

3.3   Modeling assumptions

To define the data model, we have considered the overall characteristics and operation of the DC presented in Fig. 4.4 under the following assumptions:

■ The DC is connected to a smart grid and may can act both as a consumer and a producer of energy, namely as a "prosumer". This capability is enacted by means of a local energy marketplace where stakeholders interested in trading energy can actively participate.

■ The DCs can take advantage of their energy flexibility and participate in demand response programs by changing their regular operation patterns (and therefore the electricity consumption patterns) as a response to changes, such as the electricity price variation in time or grid level flexibility requests. The DC is assumed to provide ancillary services to the local electricity distributor, when requested, such as power and voltage regulation, loss compensation, to name a few.

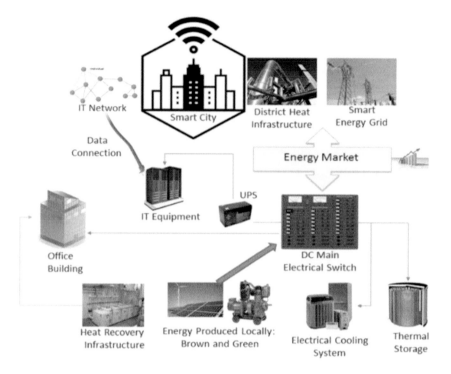

Figure 3.1: DCs energy resources and smart grid energy connections.

■ Sensors and actuators are deployed, allowing for acquiring relevant data for optimization decision making and for enforcing decisions to control the DC conditions, such as temperature, humidity, lighting, energy demand, to name a few.

■ The DC has the physical capability and infrastructure for recycling the excess thermal energy, feeding it back to the grid or heating the associated office building.

■ The DC is partially powered by renewable energy that is produced locally (collocated). This makes defining new interactions with the smart grid aiming at feeding surplus energy back to the grid possible.

■ The DC has the hardware capability to generate brown energy (i.e., diesel generators) which may be used for back-up power in case of an emergency, to reduce demand or provide energy to the grid and to provide on demand reactive power control services to the local grid.

■ The surplus of energy generated may be stored in batteries for later use. Batteries are used to guarantee power continuity during incidents and upsets, and to allow time for generators to come online. Some battery capacity may be used as energy storage to reduce or increase energy demand when requested.

At run-time, the model will be instantiated by using specific features of the modelled DC, applicable per case. Thus, only the subset of concepts relevant to the DC will be used. Being a semantic model, it uses an "open world assumption" approach, meaning that if a subset of concepts or properties is missing, the modeling and assessment processes do not fail.

We have classified the concepts describing the energy efficiency of the DC operation in two main branches: (i) *DC internal context* – defining energy-related entities and operations that are internal to the DC and (ii) *DC external context* – defining concepts allowing for DC interaction with the smart energy grid resources with the aim of increasing the energy efficiency.

3.4 DC internal data model

The semantic concepts describing the internal energy context of DC operation were defined starting from the Energy Aware Context Model (EACM) [305], which aims at representing the energy and performance features of a DC operation. EACM uses a Component – Action – Indicator based approach in which the components describe the DC computing resources which consume energy to execute workload tasks, the actions are optimized enforcing the energy-efficient execution of DC workloads, while the indicators are Green and Key Performance Indicators (GPIs/KPIs) used to characterize the efficiency of DC operations. EACM is limited on describing the energy efficiency of DC computing resources (that is, components which execute workload) ignoring other DC hardware resources (that is, cooling systems, batteries, RES, among others).

Thus, we have extended this data model to enable it to represent the energy related information of all internal components of a DC which had been classified in three main sub-ontologies (see Fig. 11.2) connected by means of partial relations: *Component Sub-Ontology*, *Energy Efficiency Optimization Action Sub-Ontology* and *Energy Efficiency Indicator Sub-Ontology*.

3.4.1 Hardware components ontology

The DC *Component* concept models identifies any active equipment inside the DC that can be monitored and controlled, and as a result it optimizes its operation. We have classified the DC Component in three main sub classes of concepts by defining and using the "has subclass" relationship (see Fig. 11.2): *components that consume energy, components that produce energy and components that store energy* (for example batteries or fuel cells).

The main characteristic modelled for <u>*components that consume energy*</u> is their energy demand value as a data property which is inherited by all modelled DC components in the hierarchical class and is determined either by means of data monitoring (if the monitoring infrastructure is available) or by means of estimation based on the component's characteristics. The components that consume energy are further classified into: (i) *non–IT components* which do not execute workload, but are used to ensure the proper conditions for workload execution or to improve the quality of personnel working conditions, and (ii) *IT components* that execute DC workloads.

The Facility and the associated Office Building are the *non-IT components* modeled for a DC (see Fig. 3.2). The *OfficeBuilding* concept encapsulates data describing the building ambient temperature and lighting, quantity of heat or surface area, and can be used for reasoning out heat transfer operations for reusing the excess DC heat in the building. The DC Facility is a generic ontology concept which specializes in the following systems:

- *CoolingSystem* – models the actual, minimum and maximum cooling capacity, input and output air temperature and quantity of heat removed. Two types of cooling systems are considered: air cooling system (*AirCooling* concept) and liquid cooling system (*LiquidCooling* concept). For each of these, relevant characteristics are defined (for example, temperature and usage %) for estimating (if the monitoring infrastructure is not available) the energy consumption overhead induced by cooling the DC.

- *HeatRecoveryInfrastructure* – describes the infrastructure (for example, heat pumps) used to recover the excess heat generated with the aim of heating the office buildings associated with the DC or passing it onto the District Heating Infrastructure. Its main function is heat absorption which measures the amount of heat recovered.

- *LightingSystem* – model properties related to the electrical lighting system of the DC, namely relative luminance, luminous efficacy (for example, how well a light source produces visible light) and covered area.

- *TransformerSystem* – deals with power conversion from AC to DC and has input and output power and the conversion loss as the its main characteristics.

The *DC IT components* are represented mainly by the computational resources which are running the client's workload. Examples of such components are servers, racks, network equipment, among others. The generic concept *ITComponent* describes their characteristics relevant to determining their energy consumption such as low and high-power states, usage levels, temperature and moisture range, and the rest. The *ITComponent* concept is specialized as an *IndividualComponent* and a *CombinedComponent*.

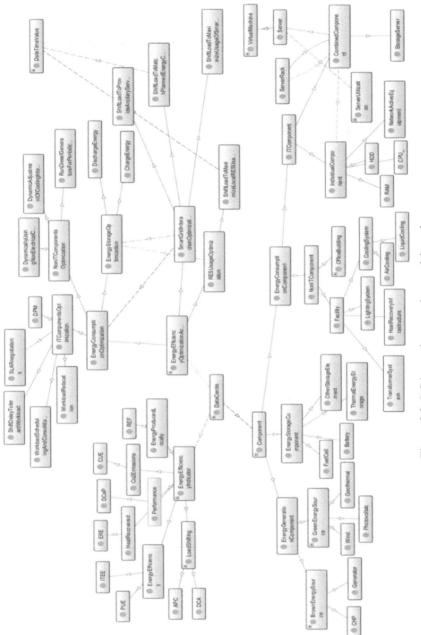

Figure 3.2: DC internal energy data model ontology.

The *IndividualComponent* concept defines the set of computational elements which are part of complex data processing systems. The following properties are considered: (i) *NetworkActiveEquipment*: usage and throughput; (ii) *RAM*: actual used capacity, total capacity, frequency; (iii) *CPU*: frequency, usage, number of cores; (iv) *HDD*: used capacity, total capacity, transfer rate.

CombinedComponent concept defines complex components that aggregate individual ones. The aggregation relation between any *IndividualComponent* and a *CombinedComponent* is expressed by means of the *hasIndividualComponent* semantic relation. Three types of such components had been modelled: *Server*, *StorageServer* and *ServerRack*. A server is modeled as being composed from memory, CPU and hard disk resources which are defined as an *IndividualComponent*. The *ServerRacks* concept is modelled using a polymorphic relation *hasIndividualComponentSU* since the Server is already defined as a *CombinedComponent*. The workload executed by each server is modelled using the *hasVirtualMachine* association relation, while for each virtual machine its relevant features modelled are the needed computational resources, namely execution time and deadlines (for delay tolerant workloads).

Table 3.1 shows the modeled energy consumption components together with their data source and type.

Components that produce energy are represented in our model by the *EnergyGenerationComponent* ontological concept which has two data type properties that will be propagated down in the inheritance tree: the maximum and actual energy production values. According to the type of energy produced the model defines (see Table 3.2):

- *GreenEnergySource* - RES components such as (i) solar panels, (ii) wind turbines and (iii) *Geothermal* convertors related to geothermal springs.

- *BrownEnergySource* - components such as *DieselGenerator*, *CHP* (Combined Heat and Power) systems.

Components that store energy have following data type properties common for all DC components that store energy: actual load and maximum charging level, discharging / charging rate and loss rate, to name a few. The model defines Battery, *FuelCell* and *ThermalEnergyStorage* (see Table 3.3) concepts for describing specific types of DC energy storage components. Additionally, for the thermal energy storage component new characteristics are modelled such as the specific heat of the substance that will absorb the heat and the temperature difference.

3.4.2 Energy efficiency actions ontology

This sub-ontology defines the class of actions that can be taken to improve the energy efficiency of DCs. Energy Consumption Optimization Actions (see

Table 3.1: Modelled DC energy consumption components.

Ontology Concept			datatype Property	Source	Value Type
IT Component			energyConsumption	Measured	double
			lowTemperatureLimit	Manual	double
			highTemperatureLimit	Manual	double
			lowMoistureLimit	Manual	double
			highMoistureLimit	Manual	double
			heatGenerationValue	Manual	double
			P_IDLE	Manual	double
			P_MAX	Manual	double
	Individual Comp				
		Network Active Equipment	throughput	Measured	double
			interfaces	Manual	int
			networkutilization	Measured	double
		CPU	clockSpeedGHZ	Measured	int
			noCores	Measured	int
			cpuUtilization	Measured	List
		RAM	totalCapacity	Measured	double
			speedMHZ	Measured	int
			usedCapacity	Measured	double
		HDD	totalCapacity	Manual	double
			RPM	Manual	int
			transferRate	Measured	double
			usedCapacity	Measured	double
	Combined Component		componentList	Manual	List
			thermicInfluence	Measured	List
			inputAirTemperature	Measured	double
			outputAirTemperature	Measured	double
			airFlow	Measured	double
		Server	serverUtilization	Measured	Map
			vmList	Measured	List
		Server Rack	ITRackDensity	Measured	double
			noServers	Measured	int
			combCompList	Measured	List
Non-IT Component			energyConsumption	Measured	double
	Facility				
		Cooling System	coolingCapacity	Measured	double
			copCoeff	Manual	List
			maxCoolingLoad	Manual	double
			minCoolingLoad	Manual	double
			inputAirTemperature	Measured	double
			outputAirTemperature	Measured	double
			heatRemoved	Manual	double
		Air Cooling	airFlow	Measured	double
			airSpecificHeat	Manual	float
		Liquid Cooling	liquidFlow	Measured	double
			liquidSpecificHeat	Manual	float
		Heat Recovery	heatAbsortion	Measured	double
		Lighting System	illuminanceValue	Manual	double
			luminousEfficacy	Manual	double
			surfaceArea	Manual	double
		Transformer System	inputPower	Measured	double
			outputPower	Measured	double
			conversionLoss	Manual	double
		Office Building	computedHeat	Manual	Map
			totalSurface	Manual	double
			ambientTemperature	Measured	double
			ambientalLinghting	Measured	double

Table 3.2: Modelled DC energy production components.

Concept		datatype Property	Source	Value Type
Green Energy Source		maxCapacity	Manual	double
		energyProduction	Measured	double
		energyProductionEfficiency	Manual	double
	Photovoltaic	totalSolarPanelArea	Manual	double
		solarPanelYield	Manual	double
		averageIrradiationOnTiltedPanels	Measured	double
		performanceRatio	Manual	double
	Wind	bladeLength	Manual	double
		windSpeed	Measured	double
		airDensity	Manual	double
		powerCoefficient	Manual	double
	Geothermal	fluidSpecificHeat	Manual	double
		flowRate	Manual	double
		sensibleHeat	Manual	double
		parasiticLosses	Manual	double
Brown Energy Source		maxCapacity	Manual	double
		energyProduction	Measured	double
		efficiency	Manual	double
		emissionsCO2	Manual	double
		sourcePriority	Manual	double
	Generator	loadTofuelConsumption	Manual	Map
	CHP			

Table 3.3: Modelled energy storage components.

Concept		datatype Property	Property	Value Type
Energy Storage Component		actualLoadedCapacity	Measured	double
		actualChargeRate	Measured	double
		actualDischargeRate	Measured	double
		maxDischargeRate	Manual	double
		maxChargeRate	Manual	double
		energyLossRate	Manual	double
		chargeLossRate	Manual	double
		dischargeLossRate	Manual	double
		maximumCapacity	Manual	double
	Battery			
	Fuel Cell			
	Thermal Storage	specificHeat	Manual	double
		temperatureDifference	Manual	double

Table 4) aim to reduce the DC energy consumption internally and generate flexible energy loads which can be then shifted to meet different smart grid level goals. EnergyConsumptionOptimization roots ontological concept models,

Table 3.4: DC energy consumption optimization actions.

Energy Consumption Optimization Actions		
Non-IT Components Operation Optimization	**IT Components Optimization**	**Energy Storage Optimization**
Adjustments of cooling intensity, allowing temperature to rise/fall within max/min limits for a given period. Reduce heat recovery/avail of free cooling to reduce cooling system electrical demand.Use non-electrical cooling to compensate the electrical one. Plan the diesel generators maintenance in accordance with grid level objectives.	Shift of delay tolerant workload from a certain period with a specific time offset. Workload consolidation at a specific time. Relocate a certain amount of workload to a partner DC. Renegotiate the customers' SLA levels for a certain period with a given percentage. Dynamic power management of components.	Charge/discharge a specific energy storage component at a specific moment of time.

through its data type property, and the three main action characteristics inherited are: (i) the energy flexibility component that will execute the action, (ii) the energy flexibility potential of the action (above and below the baseline) and (iii) the time of action execution.

RES Usage Optimization Actions aim to maximize the usage of locally produced renewable energy (collocated) in the DC operation by shifting the flexible energy load. The *RESUsageOptimization* concept defines the list of energy consumption optimization actions as data type properties with their associated execution times to modify the DC energy demand profile according to the renewable energy generation profile.

Smart Grid Interaction Optimization Actions (*SmartGridInteractionOptimization concept* in Fig. 11.2) aims to optimize the interaction of the DC with the smart grid by shifting flexible energy loads and providing on demand services to the grid. The following optimization actions are modelled:

- Adapt DC energy demand to the profile requested in Demand Response programs;

- Buy renewable energy from the local energy marketplace for immediate use;

- Buy energy from the marketplace when prices are low or sell energy (electricity or thermal) surplus to the marketplace when prices are high;

- Shift flexible energy loads aiming at providing balancing services.

The main data type properties modelled for these actions are given by the list of energy consumption optimization actions for shifting energy flexibly in the associated time frames.

3.4.3   Energy efficiency indicators ontology

DC energy efficiency indicators are defined in the model and used to evaluate the efficiency of DC operations in the context of the smart grid. The selection of the indicators represented was done in close relationship with the output of the Smart City Cluster Collaboration, composed by selected European ICT projects funded under the 7th Framework Programme, dealing with energy efficiency in DCs (see Existing Data Centres energy metrics – Task 1 [80]). The following indicators have been represented in the model, and classified per indicator category (see [329] for detailed description):

- *Energy efficiency*: Power Usage Effectiveness (PUE) – measures the DC energy consumption efficiency and IT Equipment Energy Efficiency (ITEE);

- *Energy produced locally*: Renewable Energies Factor (REF) – measures the quantity of renewable energy produced locally and used for powering the DC;

- *Energy recovered*: Energy Reuse Effectiveness (ERE) – measures the efficiency of re-using the dissipated energy from the DC;

- *Load shifting*: Adaptability Power Curve (APC) – measures how much the DC energy consumption profile has adapted to a renewable energy curve; DCAdapt (DCA) – measures how much the DC energy profile has shifted from a baseline value after the implementation of flexibility mechanisms;

- *CO2 emissions*: Carbon Usage Effectiveness (CUE);

- *Performance*: Data Centre Energy Productivity (DCeP) – overall work output of a data centre per unit of energy expended to produce it.

3.5   DC external data model

For modelling the DC external energy context, and more specifically, the interaction with the smart energy grids we had considered the DC as an Energy Element (see Fig. 3.3) of a smart neighbourhood and taken advantage of and used the Neighbourhood Information Model – NIM, developed in [257].

EnergyElement is a concept defined by NIM used to represent any entity which has a physical energy connection to the smart grid and has at least one of the

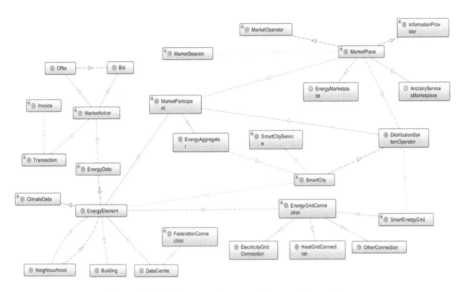

Figure 3.3: DC external energy data model ontology.

following roles: active energy consumer, active energy producer or active energy prosumer (both producer and consumer). A *Neighbourhood* is an *EnergyElement* aggregating a set of *EnergyElement* entities which are: (i) physically located in the same geographical area, (ii) willing to cooperate in buying and/or selling energy and (iii) represented as market participants in the local marketplace. We have modelled the Data Centre as an *EnergyElement* component of a *Neighbourhood* that can participate in the local energy marketplace as a prosumer of energy or as a provider of ancillary services.

We have identified the energy and climate data from the NIM relevant for characterizing the potential interactions between the DC and the smart grid. The Energy Data (see Table 3.5) describes the DC considering the quantity of energy produced or consumed and associated costs.

The Climate data (see Table 3.6) describes the weather in the location at which the DC is operated with relevance to forecasting the quantity of renewable energy produced locally.

The energy interactions between the DC and the smart grid are modelled and implemented through a Local Energy Marketplace which is characterized by:

■ timeframe describing the execution of energy transactions and may be either day-ahead, intra-day or real-time. The day-ahead value implies that market transactions are conducted one day before the actual delivery of the transacted energy products or services. The intra-day value refers to future actual delivery, as well, but taking place within the same day as the

Table 3.5: Energy Data from the NIM imported in our model.

Data Name	Value Range	Forecast Data	Source	Datatype
CO_2 emission coef.	No	No	Manual	Numeric
CO_2 emissions	No	Yes	Calculated	Numeric
Delivered Energy	No	Yes	Measured	Numeric
Energy Demand	Yes	Yes	Measured	Numeric
Energy Supply	Yes	Yes	Measured	Numeric
Produced energy	No	Yes	Measured	Numeric
RES coverage	Yes	Yes	Measured	Numeric
Energy source (wind, sun)	No	No	Manual	Enum
Energy gen. potential	Yes	Yes	Calculated	Numeric
Energy stored	Yes	Yes	Measured	Numeric
Energy storage capacity	Yes	Yes	Calculated	Numeric
Flexibility	No	No	Link	
Energy cost	No	Yes	Calculated	Numeric
Pricing data	No	Yes	Extern	Numeric
Time of Use	No	No	Extern	Text

Table 3.6: Climate Data from NIM imported in our model.

Data Name	Value Range	Forecast Data	Source	Datatype
Air temperature	Yes	Yes	Extern	Numeric
Solar irradiance	Yes	Yes	Extern	Numeric
Gust wind speed	Yes	Yes	Extern	Numeric
Rainfall total	Yes	Yes	Extern	Numeric
Reference wind speed	Yes	Yes	Extern	Numeric
Relative humidity	Yes	Yes	Extern	Numeric
Solar declination	Yes	Yes	Extern	Numeric
Wind direction	Yes	Yes	Extern	Numeric
Wind speed	Yes	Yes	Extern	Numeric

transactions. Finally, real-time value defines short time periods between the transaction execution and the actual delivery.

■ Type of service describing the subject of the transaction and may be either energy or ancillary services. The former defines the Green Energy Marketplace (GEM), while the latter defines the Ancillary Services Marketplace (ASM). The Ancillary Services Marketplace is used to rapidly mitigate potential instabilities in the energy grid, through several techniques, such as voltage/frequency regulation, loss compensation, and others.

The local energy marketplace aggregates market sessions, whose duration and time of happening vary, depending on the type of service and the timeframe of markets they refer to. During a market session, Energy Elements (for example,

DC or other prosumers) may participate by placing offers, bids and by accepting or negotiating them. The market actions are completed as energy transactions among peers. Apart from these Energy Elements, the marketplace supports other market users, as well, including:

■ Market operator which oversees the smooth operation of the Marketplace. This entity regulates the market, sets up market rules, defines market sessions and is responsible for auditing the transactions carried out in the marketplace.

■ Distribution System Operator operates and maintains the smart grid. On the GEM it is a neutral market facilitator while on the ASM it is a market participant placing bids for buying energy services when needed.

■ Energy Aggregator which is aggregating multiple energy prosumers

3.6 Model usage in data driven operations

This section presents the ways in which the defined DC energy context model was used, integrated and validated into an flexibility optimization engine [20] as support for making decisions to assess and increase the energy efficiency of a DC as an active energy resource in the smart grid.

The multistage pipeline designed to extract valuable information using the semantic model by enhancing, aggregating, filtering and processing multiple heterogeneous data sources is presented in Fig. 3.4. The defined semantic model was

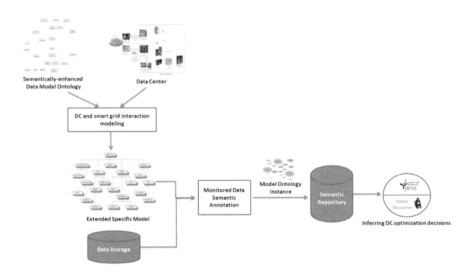

Figure 3.4: Semantic model enhanced data processing pipeline.

used to support the implementation of the following data driven operations: (1) DC and smart grid interaction modelling, (2) semantic annotation of DC monitored data and (3) inferring DC operation efficiency.

3.6.1 Modelling the DC infrastructure

Every DC has certain specificities in terms of hardware and software infrastructure, configurations and specific energy-performance dependencies that need to be captured by extending the defined semantic model. To capture the specific energy context of a DC being modelled, the ontology needs to be mapped onto the DC hardware characteristics and new concepts and relations must be defined for extending the model ontology. In this process the ontology datatype properties of the concepts, listed in Tables 3.1-3.6 as in the data source "manual", will be assigned values based on the characteristics of the DC case as described in Table 3.7.

The DC workload and energy demand was taken from the IT power consumption data logs of an operational DC [72] considering 5 minute samples normalized using the maximum power consumption of the modelled DC. The source used for energy price data is [270].

The process of generating the specific model extensions for the used case DC is detailed below. We have used Protégé ontology editing software tool [286] and OWL language [278] for defining the model concepts and relations in specific test cases using DC. Figure 5 presents how the DC cooling system concept model is created using the Protégé tool.

As it can be noticed, besides the data property assertions which are manually inserted based on the component characteristics from Table 7 there are proper-

Table 3.7: Test case DC hardware characteristics.

Component	Characteristics
Electrical Cooling System	Cooling Capacity = 4000 kWh, Minimum Cooling Load = 200 kWh, Maximum Cooling Load = 2000 kWh, COP Coefficient = 3.5
IT Computing Resources	PMAX = 325 W, Memory, Processor, Hard Drive = RAM 8 GB, CPU 2.4 GHz, HDD 1Tb
Electrical Storage System	Charge Loss Rate = 1.2, Discharge Loss Rate = 0.8, Energy Loss Rate = 0.995, Max Charge & Discharge Rate = 1000 kWh, Max Capacity = 1000 kWh
Thermal Storage System	Charge Loss Rate = 1.1, Discharge Loss Rate = 0.99, Energy Loss Rate = 0.999, Max Charge & Discharge Rate = 1000 kWh, Maximum Capacity = 3000 kWh
Diesel Gen.	Max Capacity = 3000 kWh

Figure 3.5: DC cooling component modelled in Protege.

Figure 3.6: Energy storage component modelled in Protege.

ties with default values that will be assigned with the actual monitored data fed by the monitoring infrastructure (*Measured* data annotation in Fig. 3.5). Similarly, the thermal storage can be represented as an *EnergyStorageComponent* (see Fig. 3.6).

To create a server in the model ontology the following steps are followed (see Fig. 3.7):

■ The CPU, RAM, HDD and Network Interface Card components need to be created as concepts in the model with their corresponding characteristics and inserted as *IndividualComponents* in the ontology;

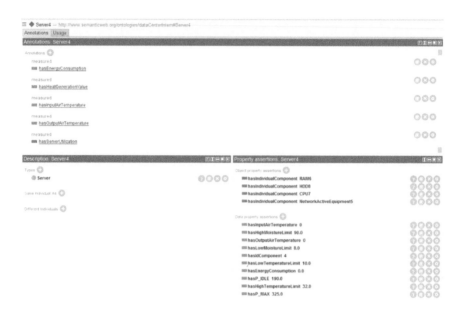

Figure 3.7: IT servers modelled in Protege.

- ■ The Server component needs to be created as a *CombinedComponent* that has the individual components defined before.

The modelling of possible actions to be executed for improving the DC energy efficiency need to be consistent with the capabilities of the hardware components that will enforce the actions. Let us consider the situation in which a peak of renewable energy is predicted at 1:00 PM. In this case, the optimization engine may decide to adjust the DC energy demand profile to the peak value of renewable energy by shifting the DC flexible energy consumption that would have occurred between 10:00 and 11:00 AM to 1:00 PM. To model such a complex optimization action first the ontology is extended with actions that will be executed by the DC components that will provide the amount of flexible energy needed:

- ■ Action 1: "Adjustment of cooling intensity with 10% between 10:00 and 11:00 AM" will be created as a member of the *DynamicAdjustmentOfCoolingIntensity* class and will be executed by the electrical cooling system;

- ■ Action 2: "Discharging the Thermal Storage with 30 kWh from 10:00 AM" will be created as an element of the *EnergyStorageOptimization/DischargeEnergy* class and will be executed by the thermal storage system;

- ■ Action 3: "Charge the Thermal Aware Storage with 30 kWh from 1:00 PM" will be created as part of the *EnergyStorageOptimiza-*

Figure 3.8: Optimization action modelled in Protege.

tion/ChargeEnergy class and will be executed by the thermal storage system.

After the above optimization actions individuals have been created a complex action of "Shifting 30 kWh of flexible energy from 10:00 am to 1:00 PM" will be carried out and placed under the *RESUsageOptimization* concept. The complex optimization action will have the three actions defined above in the related list (see Fig. 3.8).

3.6.2   *Semantic annotation of monitored data*

After the DC energy data model is extended and customized with the specific characteristics of the used DC case, it may be used for semantically annotating the data acquired by the monitoring infrastructure. A new instance of the ontology will be defined and stored each time new data is fed by the monitoring infrastructure. More specifically, the datatype properties of the ontology concepts listed in Tables 1-6 as having the data source "monitored" will be assigned values based on the current (latest) data provided by the monitoring infrastructure.

Due to the large number of ontology individuals needed to be instantiated, updated or classified in the defined taxonomy, this process must be automated. Also, we need to assure a semantic check and automate reasoning on the energy data which cannot be done by using only the relational data. Our solution is to monitor the DC hardware components, used to store the monitored data coming from sensors in DC Monitored Data Storage in a relational database (see Fig. 3.9 for DB schema) and to map the DC energy context model ontology concepts onto the database tables obtaining thus keeping related instances updated. This allows us to keep the context model concepts in memory, reasoning relevant to the model concepts and their instances of occurrence in the relational database.

To connect the defined semantic data model with the monitored energy data stored in the relational database various alternatives had been investigated considering the availability of a reasoner for inferring the DC operational state and of a specialized query language where considered [256], [79], [78]. We have chosen

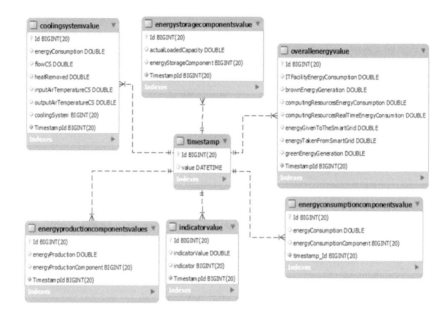

Figure 3.9: Monitored data storage schema.

to use D2RQ [88] due to the fact that classes, object properties and data properties can be managed in the form of database tables, and there is a clear separation between the ontology data (individuals) and ontology structure, which can be easily integrated with Jena [181] a powerful tool for reasoning on ontologies. Figure 11.10 shows how our data model ontology is integrated with the energy database to query and reason about DC operational efficiency.

Individual energy data is stored in the SQL DC Monitored Data Storage. The relations between the ontology specific elements (for example, ontology classes, data properties, object properties, and subclass relations) and the database tables are defined in the Mapping File. Reasoning and query processes are defined considering DC Semantic Energy Context Model ontology as well as the mapped data from the monitored database and are used to assess the efficiency of DC energy operation. As depicted in Fig.3.10, the Server Model is annotated with *@OntologyEntity* marking the Java class as the model mapped to the Server ontological individual. Further annotations like *@ObjectProperty*, *@DynamicField*, *@InstanceIdentifier* and *@OntologyIgnore*, offer support for more complex data mapping and annotation scenarios (see Fig. 3.11 to see how a list of *VirtualMachine* instances are linked to the Server class as an object property)

The annotation of the stored data individuals is done as depicted in Fig. 11. A server repository is created by extending the OntologyRepository class, the first generic type specifying the entity to which it is linked to, and the second generic

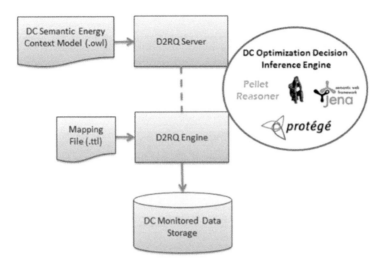

Figure 3.10: DC Monitored data semantic annotation software infrastructure.

```
// a) Server Model
@OntologyEntity
public class Server extends CombinedComponent {

    private String serialNumber;
    @DynamicField
    protected double energyConsumption;
    @ObjectProperty(value = "hasVirtualMachines", range = VirtualMachine.class)
    private List<VirtualMachine> virtualMachines;
    ...


// b) Server Repository
public class ServerRepository extends OntologyRepository<Server, Long> {
}


// c) Access Server Resources
ServerRepository serverRepository = new ServerRepository();
List<Server> server = serverRepository.findAll();
...
```

Figure 3.11: Example of DC monitored data semantic annotation in case of Server ontological concept.

type specifying the ontology individual identifier. By using an instance of the ServerRepository class, one can easily have access to the basic CRUD operations over the Server individuals.

3.6.3   Assessing the DC operation efficiency

Semantic reasoning and querying techniques will be defined and used on the defined semantically enhanced data model instances with the goal of: (i) evaluating if energy efficiency policies defined for the DC hold or not for the current ontology instance, (ii) inferring new DC energy efficiency knowledge and complex energy dependencies, (iii) estimating the energy budget of the DC and (iv) checking consistency and validation.

The reasoning techniques employed on the defined semantic model assume the following sequence of operations: (i) defining reasoning and query rules in SWRL (Semantic Web Rule Language) [334] and SPARQL (recursive acronym for SPARQL Protocol and RDF Query Language) [333] languages, (ii) injecting the defined rules of the model ontology and (iii) evaluating the injected rules by means of reasoning engines.

Table 3.8 presents examples of reasoning and querying rules which can be evaluated on the defined semantic model. For rule evaluation we have used Pellet reasoner [283] along with Jena. The advantage brought about by reasoning over ontologies is that a reasoning rule that refers to a semantic model concept instance having a property which has no assigned value (the information is missing from the ontology because it was not fed by the monitoring infrastructure), is still executed and some information is inferred.

3.7   Conclusions

In this chapter we have presented the design and implementation of a semantically enhanced data model for representing and analysing the energy characteristics of a DC as well as the reasoning-based methods used to assess the energy efficiency of the DC operation for optimal smart energy grid integration. The model design has been driven by an effort to identify the main concepts and relations describing the energy efficiency of a data centre operation, and by implementing them in an ontological manner. To describe the efficiency of the DC operations in the context of the smart grid *concepts describing the DC external energy and business context* have been defined and implemented. Finally, the chapter shows how the model can be used to support data driven decision making operations to increase the energy efficiency of a test case DC. For this a pipeline of software tools and operations have been proposed addressing the DC infrastructure modelling, semantic annotation of monitored data and the reasoning and inference of DC optimization actions.

Table 3.8: Reasoning and querying rules examples.

Objective	Inference Rule Implementation (SWRL/SPARQL)
Infer the energy consumption of the cooling system to cool down the DC servers used to run the workload	EFF= $\sum \frac{P_{IDLE} + ServerUtilization * (P_{MAX} - P_{IDLE})}{COP_1}$ *SELECT*(SUM ((xsd:double(?P_MAX) - xsd:double(?P_IDLE))?serverUtilization + ?P_IDLE) / (?t*?t*?c1 + ?t*?c2 + ?c3))AS ?SUM) *WHERE* {?r rdf:type dci:Server . ?r dci:hasP_MAX ?P_MAX . ?r dci:has P_IDLE ?P_IDLE . ?r dci:hasServerUtilization ?serverUtilization . ?ac rdf:type dci:AirCooling . ?ac dci:hasCopCoeff1 ?c1} *GROUP BY* ?ac
Infer if the optimal load level defined for the DC servers holds	$0.6 \leq ServerUtilization \leq 0.8$ *SELECT* Server (?x), hasServerUtilization(?x, ?u), greaterThan(?u, 0.6), lessThan(?u, 0.8) - > hasOptimalLoadLevel(?x, true)
Estimate the DC energy budget out of the current semantic data model instances	*SELECT* ?aitp ?pr ?spy ?tspa ?alc (SUM (((xsd:double (?P_MAX) - xsd:double (?P_IDLE)) * ? serverUtilization + ? P_IDLE) / ((?t*?t*?c1 + ?t*?c2 + ?c3)) AS ?SUM1) SUM (((xsd:double(?P_MAX) - xsd:double(?P_IDLE)) * ?serverUtilization + ?P_IDLE)) AS ?SUM2) (?aitp * ?pr * ?spy * ?tspa + ?alc + ?SUM1 + ?SUM2 AS ?RESULT) *WHERE* { ?r rdf:type dci:Server ?r dci:has P_MAX ?P_MAX ?r dci:has P_IDLE ?P_IDLE ?r dci:hasServerUtilization ?serverUtilization . ?ac rdf:type dci:AirCooling . ac dci:hasCopCoeff1 ?c1 . ?ac dci:hasCopCoeff2 ?c2 . ?ac dci:hasCopCoeff3 ?c3. ?pv rdf:type dci:Photovoltaic . ?pv dci:hasAverageIrradiationOnTiltedPanels ?aitp . ?pv dci:hasPerformanceRatio ?pr . ?pv dci:hasSolarPanelYield ?spy . ?pv dci:hasTotalSolarPanelArea ?tspa . ?tes rdf:type dci:ThermalEnergyStorage . ?tes dci:hasActualLoadedCapacity ?alc } *GROUP BY* ?ac ?aitp ?pr ?spy ?tspa ?alc
Estimate the impact of the dynamic adjustment of cooling intensity over and above the DC energy budget	*SELECT* ?adjust (SUM(((xsd:double(?P_MAX) - - xsd:double(?P_IDLE)) * ?serverUtilization + ?P_IDLE) / ((?t*?t*?c1 + ?t*?c2 + ?c3)) AS ?SUM) (?adjust / 100.0 * ?SUM AS ?RESULT) *WHERE* { ?r rdf:type dci:Server . ?r dci:hasP_MAX ?P_MAX . ?r dci:hasP_IDLE ?P_IDLE . ?r dci:hasServerUtilization ?serverUtilization . ?ac rdf:type dci:AirCooling . ?ac dci:hasCopCoeff1 ?c1 . ?ac dci:hasCopCoeff2 ?c2 . ?ac dci:hasCopCoeff3 ?c3 . ?action rdf:type dci:DynamicAdjustmentOfCoolingIntensity . ?action dci:hasCoolingIntensityLevelAdjust ?adjust } *GROUP BY* ?ac ?adjust

Chapter 4

Managing the Safety in Smart Buildings using Semantically-Enriched BIM and Occupancy Data Approach

Muhammad Arslan, * *Christophe Cruz* and *Dominique Ginhac*

4.1 Introduction

Buildings are one of the critical components of a human's living environment [183]. The idea of smart buildings initiated with an increase in the utilization of advanced Information and Communication Technologies (ICTs) for buildings and their linked systems (for example, Heating, Ventilation, and Air Conditioning (HVAC)) so that they and their occupants can be monitored and controlled remotely, in a cost-effective way. The integration of ICTs with building systems is the prerequisite for achieving the realization of smart buildings. This includes deploying sensors in buildings for monitoring occupant behaviors, processing and enriching collected occupants' data, using state-of-the-art big data methods,

Laboratoire d'Informatique de Bourgogne (LIB) - EA 7534, Univ. Bourgogne Franche-Comte, Dijon, France.
* Corresponding author: muhammad.arslan@u-bourgogne.fr

cloud and fog computing for storing the acquired behaviors and performing different real-time data analytics to mention a few [183]. In the past decades, a huge number of studies have been conducted on smart buildings by incorporating the dynamicity of occupant behaviors for providing reliability, safety, and energy efficiency in building services to occupants without compromising their comfort level [33, 34]. While executing responses to occupant actions permits building supervisors and (H&S) managers to investigate their complete effects in the building environment. However, this process of analyzing behaviors using real-time sensory data gets complex when the dynamicity (spatial and semantic changes) of the building facilities is included for understanding occupant behavior. The building facilities are dynamic as they evolve spatially as well as contextually [33, 34]. Here, a context is any type of data that defines the building facilities. An ideal example of a dynamic building environment that evolves in both geometry and contextual information is the construction site. New structural and architectural supports are frequently added on worksites, while others are detached based on the construction work needs. However, constructed facilities undergo less geometrical evolutions but many changes in the contextual information related to building facilities [37]. For instance, an office in the building is now a storage and has a different purpose that is a semantic change. The spatial and semantic evolutions open more data management challenges to keep track of the relevant changes in the attributes of a building environment. Existing literature consists of many building monitoring solutions based on different ICTs for occupant behavior analysis [234, 54]. These systems are executed in real-time for sensory data collection and give timely notifications using handheld devices to building supervisors and safety managers to take necessary actions in case of any abnormality observed in the occupant behaviors across building facilities. Existing literature suggests that, for studying the occupant behaviors, occupancy is the fundamental pre-requisite to infer different occupant activities [368, 383, 70, 164]. For monitoring the building occupancy, there exist two methods for data collection which are; direct and indirect. The 'direct' method collects occupancy data using different ICTs (for example, smart cameras, Radio Frequency Identification (RFID), beacons, cellular networks, to name a few). Whereas, the indirect methods compute occupancy based on the data acquired using environmental monitoring sensors for example, Carbon dioxide levels and stochastic modeling. To have more information on the methods for occupancy data collection, see the research in [390, 395]. Once the occupancy data is collected, it is enriched with the contextual information related to the building. The building information is typically extracted using an OpenStreetMap (OSM) or BIM-based data files [33]. However, the BIM approach is recommended as it holds up-to-date building information throughout the building lifecycle [33]. Existing literature contains numerous approaches based on building information and real-time sensory data for generating semantically-enriched BIM models to study occupancy behaviors for different safety management applications (as mentioned in Fig. 4.2). However, the dynamicity of building while understanding the occupancy levels in a building for ensuring safety facilities is not yet adequately explored in the existing literature. To address this gap, this study utilizes the framework OBiDE to model building occupancy for designing a safety management application in a dynamic building environment scenario [36]. The rest of the paper is sequenced as follows: Section 4.2 describes the background literature of the presented study. Section 4.3 presents proposed system implementation in which the occupancy is measured using the spatio-temporal trajectories of occupants which were collected in a building using Bluetooth Low Energy (BLE) beacons. Secondly, the occupancy data is integrated with BIM software. Section 4.4 provides a discussion and Section 5.6 presents a conclusion and some limitations of this study.

4.2 Background

Occupants perceive, move and perform actions in the buildings as well as interact with the building environment (other occupants and the building objects) for achieving desired satisfaction [368, 383, 70, 164]. The occupant actions are typically initiated from a series of entangled mental thoughts which control their actions and interactions with the building environment. These observable actions or interactions in response to external or internal stimuli are called occupant behaviors [368, 383]. The understanding of occupant behaviors implies inferring the drivers and goals using their actions to identify their influence on the environment and thus to decide which of them are more suitable for effective building management. Responding to the occupant behaviors during their execution permits building analysts to study their effects on its environment [368, 383]. However, the consequences of their behaviors may result in serious problems while maintaining safety in buildings. To avoid such undesirable situations, a dynamic and real-time understanding of occupant behaviors is required while their actions are still evolving in the changing environment.

Over the last decade, a strong interest in understanding occupant behavior has attracted researchers for better monitoring and control of the building facilities. To understand the occupant behaviors for different building management applications, in the existing literature [368, 383, 70, 164], an extensive range of different types of sensors (wired and wireless) are present for monitoring occupants and the environment. These sensors contribute towards the enrichment of environmental information for modeling occupant behaviors and their interactions (energy consumption, amongst other examples as well) with the buildings. After collecting the required sensory data, this data is mapped with the corresponding building information using a building context to analyze the occupant behaviors [384, 33]. To utilize the building information, Building Information Modeling (BIM)-based platforms are use and preferred over traditional 3D Computer-aided design (CAD)-based systems in the AEC industry [34]. A BIM model is a digital representation of the physical and functional characteristics of a facility [36]. The BIM model provides a centralized source of shared knowledge of facility information for the building managers to use and maintain throughout its lifecycle [92]. Each building component in a BIM model is a solid object containing its physical as well as the semantic information (that is, alphanumeric properties) [92]. However, BIM models lack the feature of evaluating the current state of the building environment in real-time for building management processes [336]. Here, a state of a building object (for example, a building facility) refers to its environmental data for example, temperature, humidity, carbon dioxide among others the status of people inside the location (for example, tracking, monitoring occupancy and activity identification), and the analysis of the utilization rate or location of building equipment and performing localization and preventive maintenance of building objects for different BIM-based application scenarios [346]. The process adding meaningful information about states of building objects in a digital BIM model is referred to as semantic enrichment [303]. However, the resulting building model having undergone the semantic enrichment processes is referred to as a semantically-enriched BIM model. Also, for designing real-time building monitoring solutions using the BIM models, building objects (locations) need to be constantly updated with the current state of the actual building environment [336, 346]. To incorporate the real-time sensory data of a building environment into a BIM model for studying occupant behaviors in the context of a building facility, wireless sensor technology has gained great importance in real-time monitoring of the buildings [383]. "Some of the latest BIM and sensor-based integrated solutions for different building management applications from the literature are presented in Table 4.2". As it can be observed, the integration of BIM data with real-time sensory data acquired from different sensors has provided us with a centralized digital building model to study different types of occupant behaviors for building management applications [346]. However, the quality of acquired sensory data for the enrichment of BIM models differs significantly based on the resolution of the deployed sensors "see Fig. 4.1".

The resolutions concerning spatial, temporal, occupant and semantic information level are combined to determine the overall system resolution for capturing the occupancy data to study different occupant behaviors [368, 383, 251, 344]. The spatial resolution defines the size of the targeted building area (for example, buildings, floor, rooms) for data collection. Whereas occupant resolution typically has 4 different levels of information which are; (1) occupancy detection using 0 or 1 values, (2) counting occupants in a building facility, (3) recognizing the occupant identities using their information profile, and (4) distinguishing the occupants' activities based on the use case application. In addition, temporal resolution defines the smallest time period in which variations in spatial and occupant resolutions are reported by the installed sensors across building facilities

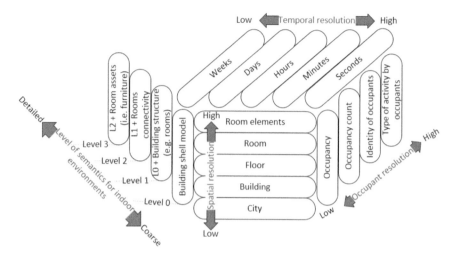

Figure 4.1: System resolution (Modified from [37]).

[37]. Finally, the semantic resolution describes the level of detail of contextual information of a facility extracted from a BIM model which is mapped with the spatial, temporal and occupant data. As the system resolution of sensors increases, the building facilities get smaller, the information linked to the building locations gets more precise, occupants get more recognizable individually and hence, their activities can be studied to infer their behaviors for different applications [37]. However, the process of understanding occupant behaviors gets more complex when the dynamicity of the environments is built into the understanding process [87]. The change in the purpose or the position of the locations in the buildings will result in different occupant behaviors [383]. The updated location context with the previous contextual information is required to be captured for studying the behaviors of the occupants in detail in relation to the changes that have occurred in the building. The historicization of building information in the form of a semantic BIM model enriched with the occupancy data will help building managers for conducting cause and effect analysis [34].

4.3 Proposed system

The six different processes "see Fig. 4.3" are used to develop the OBiDE framework "see Fig. 4.4" which is later used to design a prototype system application "see Fig. 4.5" which is based on;

No.	Use case	Data type	Key technologies / concepts
1	**BIM-based building fire emergency management** (Ma and Wu 2020)	Temperature, smoke and carbon dioxide data	Autodesk Revit software, SQL server and voice broadcast device.
2	**Stations-oriented indoor localization** (Yoo et al. 2019)	Wi-Fi signal strength	Dynamo, Autodesk Revit software and MySQL database.
3	**Infrastructure-free visual indoor localization approach** (Acharya et al. 2019)	Images	Cameras
4	**Wireless electric appliance control for smart buildings** (Rashid et al. 2019)	Location data	Ultra-wideband (UWB) anchors and real-time location tracking system (RTLS)
5	**A cyber-physical system approach for building efficiency monitoring** (Bonci et al. 2019)	Temperature and carbon dioxide data	Matlab, Autodesk Revit software and SQL Server
6	**Augmented reality system for facility management** (Baek et al. 2019)	Location data	Nvidia graphics processing unit (GPU)
7	**Detecting, identifying and localizing lighting elements in a building** (Troncoso-Pastoriza et al. 2019)	Lighting data	LiDAR sensors and an inertial measurement unit (IMU)
8	**Optimal path planning for dynamic building fire rescue operations** (Chou et al. 2019)	Location data	Bluetooth sensors
9	**Indoor thermal environmental design system** (Fukuda et al. 2019)	Temperature and humidity data	Autodesk Revit software and Augmented Reality (AR) devices
10	**BIM-sensor integration for responsive building management and operations** (Kazado et al. 2019)	Temperature and carbon dioxide data	Autodesk Revit software and Autodesk Navisworks software
11	**Integrating sensor, occupant and BIM data for building performance management** (Rogage et al. 2018)	Temperature and humidity data	SQL Server
12	**Building environmental monitoring under BIM environment** (Zhong et al. 2018)	Temperature, humidity, noise, carbon monoxide, carbon dioxide and metha|	SPARQL Protocol and RDF Query Language
13	**An IoT visualization BIM platform for decision support** (Chang et al. 2018)	Temperature, humidity and carbon dioxide data	Arduino, Raspberry Pi and Dynamo
14	**BIM-based indoor guidance system** (Ferreira et al. 2018)	Location data	Bluetooth Low Energy (BLE) beacons
15	**BIM-based system for building performance monitoring** (Kang et al. 2018)	Temperature, humidity and light data	DHT11 sensors and PT550 light sensors
16	**Government Open data and sensing data integration** (Lee et al. 2019)	Vibration and temperature data	MariaDB and Autodesk Revit software
17	**Real-time dust monitoring and visualization in BIM** (Smaoui et al. 2018)	Dust data	PMS5003 and PMS7003 dust sensors

Figure 4.2: BIM and sensor-based existing systems for facility management applications.

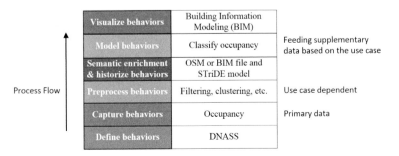

Figure 4.3: An OBiDE framework having 6 different processes.

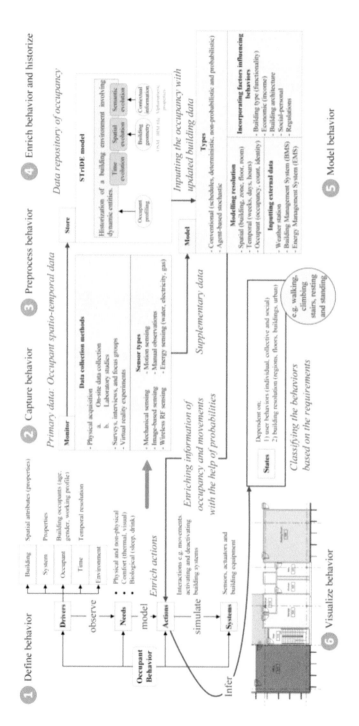

Figure 4.4: Generalized OBiDE framework (Modified from [34]).

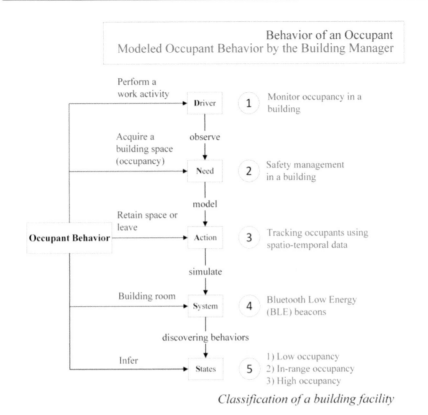

Figure 4.5: Occupancy modelling use case using DNASS.

4.3.1 Capturing the occupancy data

As occupancy is the prerequisite for any type of occupant behavioral understanding, an OBiDE framework is based on capturing the spatio-temporal data of occupants in dynamic building environments. Around 200 BLE beacons "see Fig. 4.6" were mounted inside a university building "see Fig. 4.7". Due to the environmental conditions, if the beacons are relocated in the dynamic building facility, the positions of the beacons in the system to tag the building locations against each longitude and latitude pair value need to be updated accordingly. To acquire the location coordinates (longitude and latitude pair values) of occupants, an Android-based application is used in the smartphone devices of the occupants. As an application launches in a device, it senses the neighboring beacons and picks the best three beacon signals for performing the geo-localization. Based on the Received Signal Strength Indicator (RSSI) of the beacons "see Fig. 4.8", location coordinates are generated and stored in a document database, i.e., Mongodb for further processing in the R studio. The collected dataset contains 8426 spatio-temporal points of 11 occupants having a sampling rate of 5 seconds during different times in an interval of 12 days. The dataset and its description can be found in [35].

Figure 4.6: Bluetooth beacons.

Figure 4.7: University building.

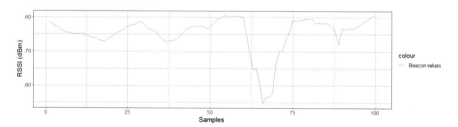

Figure 4.8: BLE beacon RSSI strength.

4.3.2   Preprocessing the occupancy data

Once the spatio-temporal data of occupants is captured in the form of trajectories, it is filtered and preprocessed based on the used case application. Spatio-temporal data of occupants often suffers from noise and environmental interferences [396, 385]. To reduce the level of noise in the collected data, median filtering is performed as it is preferred for datasets having outliers with low deviations [396]. "A processed trajectory sample is shown in Fig. 4.9".

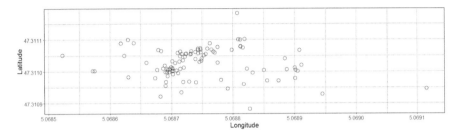

Figure 4.9: Trajectory sample of collected BLE data.

4.3.3   Semantic enrichment and storing the occupancy data

After preprocessing the occupancy data, the data is contextually enriched with the corresponding building locations to add more meaning to the collected trajectories using the building information extracted from BIM and an OSM file "see Figs. 4.10 and 4.11".

SPARQL Results (returned in 35 ms)

Trajectory	UserName	Location
⎘ dm: TrajUser1-1	⎘ User 1	⎘ Corridor1
⎘ dm: TrajUser1-2	⎘ User 1	⎘ Storage room2
⎘ dm: TrajUser1-3	⎘ User 1	⎘ Office1
⎘ dm: TrajUser1-4	⎘ User 1	⎘ Outdoor pathway
⎘ dm: TrajUser1-5	⎘ User 1	⎘ Office2

Figure 4.10: Semantically-enriched spatio-temporal trajectory of an occupant having multiple timeslices.

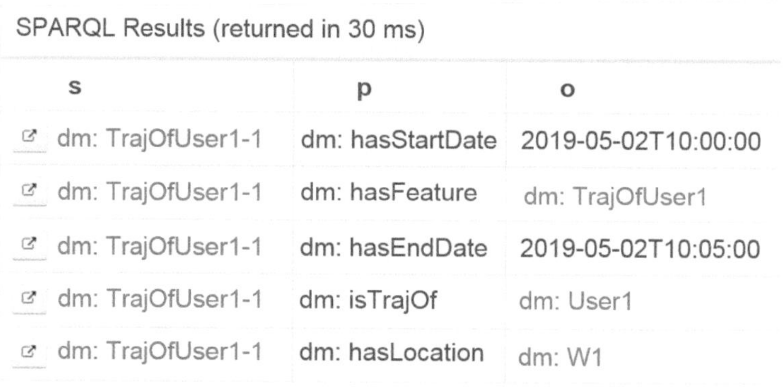

SPARQL Results (returned in 30 ms)

s	p	o
☑ dm: TrajOfUser1-1	dm: hasStartDate	2019-05-02T10:00:00
☑ dm: TrajOfUser1-1	dm: hasFeature	dm: TrajOfUser1
☑ dm: TrajOfUser1-1	dm: hasEndDate	2019-05-02T10:05:00
☑ dm: TrajOfUser1-1	dm: isTrajOf	dm: User1
☑ dm: TrajOfUser1-1	dm: hasLocation	dm: W1

Figure 4.11: The description of a timeslice from an occupant trajectory.

"Table 4.1 describes the mapping information for trajectories". Later, these spatio-temporal movements are stored in the Semantic Trajectories in Dynamic Environments (STriDE) data model. To have more details on the STriDE model, see the research of [87].

Table 4.1: Mapping of BLE data with the building information.

Location coordinates [Lon, Lat]	Geometry type	Floor	Room ID	Building Name
{[5.0685805,47.310967], [5.0683771,47.310974], [5.0688146,47.311075], [5.0685677,47.311256], [5.0685651,47.311254]}	Polygon	2	Office2	IUT

4.3.4 Modeling occupancy behaviors

Based on the used case application requirements, occupant behavior modeling techniques are applied to the processed spatio-temporal occupancy data along with their supplementary data and the building environment for achieving the desired classification of behaviors. There exist probabilistic and non-probabilistic approaches in the literature [368, 383, 70, 164] for occupant behavior modeling. To show a proof-of-concept application of modeling occupancy behaviors, the Hidden Markov Model (HMM)-based probabilistic method is chosen. The HMM describes the occupant trajectories which have been captured from the BLE beacons as a series of Markovian stochastic processes where the probability distribution of a future state (for example, low, in-range and high occupancy) of a stochastic (that is, a random) process (a spatio-temporal processed occupant trajectory) at a user's building location exclusively depends on its present occupancy state. This process of inferring the occupancy states using HMM disregards the requirements of incorporating the historical states and eventually minimal training data is required for modeling using it.

Generally, an HMM (λ) is defined using three parameters which are, transition probability matrix (A), emission probability matrix (B) and initial state probabilities (π). To train the HMMs for each building location, the Baum-welch algorithm is used which tunes the model parameters (A, B, π) for maximizing the probability of the observation sequences given by the model. The initial probabilities of the hidden states which are; 1) low occupancy, 2) in range occupancy and 3) high occupancy are assigned an equal probability of occurrence that is, $\pi = \{\pi_1, \pi_2, \pi_3\} = \{1/3, 1/3, 1/3\}$. For configuring the 'in-range occupancy' and 'high occupancy' states' values, the CLG Guide for Fire Safety Risk Assessment in Small and Medium Places of Assembly [1] is used. This Guide entails a method to keep the number of occupants in a building optimal to ensure safety in buildings. It defines a formula (that is, number of occupants = Area of building floor (m²)/Occupant density) to calculate the number of occupants who can safely reside in the building by considering its functional attributes. To input the occupant density, the Guide mentions the occupant density in the case of offices is 6.0m²/occupant. The working example is explained in Fig. 4.12.

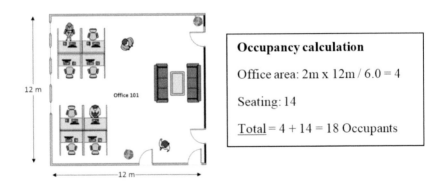

Figure 4.12: Calculating the occupant density in a building location.

For a more comprehensive overview of the literature on occupant density, the reader is referred to [1]. After training the HMMs for each building location, occupant trajectories are categorized into three states which are S_1, S_2 and S_3 using the Viterbi algorithm. The Viterbi algorithm generates the most probable sequence of states which have produced the observations under the conditioned HMM model by implementing a global decoding mechanism. "As an example, Fig. 4.13 shows the categorization of a trajectory sequence of an occupant".

Figure 4.13: Decoded sequence states S_1, S_2 and S_3 of an occupant trajectory.

4.3.5 Visualizing the classified occupancy behaviors

After categorizing the occupancy states, the latest occupancy states "see Fig. 4.13" are imported into an open-sourced BIM software that is, Autodesk Revit. The BIM model and its objects (building locations) are enriched with their corresponding states and occupancy information. For generating the occupancy visualizations, like a Revit plug-in 'Dynamo' is used. Dynamo offers to program different pieces of logic with the help of a graph by connecting "nodes" with "wires" using a visual interface by reducing the need for extensive coding. More information on Dynamo can be found in its user manual [338]. The Dynamo graph is constructed to extract the building locations from a model which were labeled as 'rooms'. Later, this list of building rooms is enriched with their corresponding occupancy states for generating visualizations on a BIM software "see Fig. 4.14". The color scheme designed for the occupancy visualizations on a BIM model is for the sake of demonstrating the functionality of the developed system by showing the building locations by changing their colors. However, it can be changed based on user preferences.

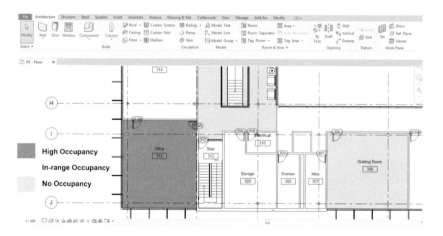

Figure 4.14: Visualizing occupancy on a building model in Revit software.

4.4 Discussion

The main research question which guided this research was, how to establish an understanding of the occupancy in a dynamic building environment The research question was motivated by the fact that occupant behaviors occupancy as in this case evolves with the buildings. Existing systems for occupant behavioral understanding do not incorporate the information of dynamic environments where the building objects (occupants and building locations) evolve geometrically or in relation to semantic information. The change in the semantics of building locations occurs more often in constructed facilities whereas, the spatial changes take place in a building life cycle rarely. Such changes in the building need to be incorporated in the occupancy behavior understanding process as a change in the purpose or a position of building locations will result in different occupancy behaviors which ultimately represent different occupant activities. For instance, an occupant in a room labeled 'Office' is identified as a worker. Whereas, over time, due to the change in the purpose of the location, that 'Office' is now a 'Storage room'. The system should identify the same occupant as an 'intruder' if he does not have access to this location. However this will not happen as the system cannot historicize the details of semantic evolutions of a building environment. The updated spatial and semantic information related to the building locations with the previous information will contribute towards improved occupant behavioral understanding relating to different changes which occurred in the building. To address this requirement of monitoring occupancy in dynamic building environments that contain evolving building objects, a prototype system was designed which; (1) defined the required occupant behaviors using Drivers, Needs, Actions, Systems and States (DNASS) ontology, (2) captured spatio-temporal data of occupants using BLE beacons (3) preprocessed the occupancy data, (4) performed semantic enrichment to add building context to occupancy data and storing it into the STriDE model to maintain the historicization of occupancy in the context of a building environment, (5) categorized the occupancy into 3 levels (low occupancy, in-range occupancy, and high occupancy) using a HMMs-based probabilistic model and (6) visualizing the occupancy using BIM software.

For studying the occupancy behaviors, BLE beacons were chosen for spatio-temporal data acquisition. The choice of BLE beacons was made because of its low deployment cost and availability for the application. Based on the proximity analysis done by the BLE beacons, locations of occupants were specified as longitude and latitude pair values captured and processed. The BLE beacons have provided the proximity to the occupants in a building. However, the proximity is not the same as the exact locations of occupants (that is, absolute values of geographical coordinates). For developing the prototype system, a high precision in the occupants' proximity was not required as trajectory data was needed for detecting the occupancy at the room level rather than capturing the exact positions of occupants inside rooms. Later, the occupancy behaviors were modeled using

Table 4.2: Potential applications of occupancy monitoring in Building management [2].

No.	Building management areas	Potential use cases of occupancy monitoring
1	Health and Safety (H&S)	Occupancy data providing the occupants' presence can be used by the emergency crew members to identify the occupant locations in case of an accident.
2	Operation and Maintenance (O&M)	Occupancy data indicating which building facilities have been used and therefore may need maintenance. Also, maintaining the building resources based on the occupancy of the building spaces.
3	Security	Triggering the alerts to building supervisors and H&S safety managers in case of unauthorized personnel occupancy in restricted areas of a buildings.

HMMs to categorize the occupancy of a building into 3 levels (low occupancy, in-range occupancy, and high occupancy). The BIM-based visualization is generated after enriching the BIM model with the occupancy information, which can help building managers to quickly analyze occupancy levels of different rooms not only for ensuring safety in a building but also to support them for optimizing the operational efficiency as mentioned in Table 4.2. Also, the developed system will allow the identification of occupants present across rooms and assist in analyzing the historical room's utilization over time.

4.5 Conclusion

Occupant behavior is an important factor in attaining building performance goals that revolve around H&S, O&M and security. The implementation of smart buildings has integrated building systems using different sensors and actuators for optimizing the building performance based on occupant behaviors. However, if the occupant behaviors, occupancy in our case, captured from the sensory data are not adequately mapped with the updated contextual information of a building. This leads to inefficient utilization of building resources and eventually degrades the overall building performance. To study the occupant behaviors by enabling the historicization of spatial as well as contextual evaluations of a building environment, a prototype system is presented for analyzing occupancy behaviors using BIM software. One of the limitations of the developed system is the fact that the study is executed in an already constructed building and involves a limited amount of semantic information about a facility for a BIM model enrichment. Additional work is required to include more semantic information from different data sources and to empirically test the system application on the dynamic building environments where spatial and contextual evolutions occur more often. This will help to understand the occupancy across different building areas during different times of the day. On the occurrence of an accident or an unauthorized personnel occupancy, alerts can be generated to building managers to reduce potential safety accidents. For the future direction of this research, it needs to be mentioned that currently this research is solely based on ICT concepts which were utilized based on the spatio-temporal data processing for occupancy monitoring application. This study presents a baseline model to monitor occupancy which is regarded as the prerequisite of occupant behavior understanding. To infer occupant activities in the building environment, a multidisciplinary approach is needed for developing advanced behavioral models. This can be achieved by combining the knowledge of occupancy-related behavioral, social and psychological factors other than those which contribute to occupant behaviors over time.

Acknowledgments

The authors thank the Conseil Régional de Bourgogne-Franche-Comté and the French government for their funding. The authors also want to especially thank the director of the l'IUT de Dijon-Auxerre for allowing us to host the living lab and Orval Touitou for his technical assistance to this research work.

Chapter 5

Belief Rule-Based Adaptive Particle Swarm Optimization

Mohammad Newaj Jamil,[a], Mohammad Shahadat Hossain,[a] Raihan Ul Islam[b]* and *Karl Andersson[b]*

5.1 Introduction

Particle Swarm Optimization (PSO) is one of the population-based iterative evolutionary algorithms, which is inspired by the food searching activities of a flock of birds [393]. The success of an evolutionary algorithm depends on the balance between exploration and exploitation in the search space. The evolutionary algorithm is an example of a meta-heuristic algorithm, where the two main components are exploitation and exploration. Exploitation is the process of generating a new solution using the information from the current focus area of the search space and directing the search around a better candidate solution to find the optimum correctly. On the other hand, exploration is the process of exploring the search space more to examine various regions in the problem space in order to find a good global optimum, which assists to refrain from being trapped in local optima.

[a] Department of Computer Science & Engineering, University of Chittagong, Chittagong.

[b] Pervasive and Mobile Computing Laboratory, Luleå University of Technology.

* Corresponding author: hridoyjamil10@gmail.com

The PSO algorithm has three parameters, namely inertia weight (w), cognitive factor (c_1), and social factor (c_2). These parameters significantly influence the performance of the PSO algorithm. Besides, it often suffers from the problem of getting stuck in local optima [219].

In PSO, the inertia weight is used to control the influence of the previous history of velocities on the current velocity and balancing the global and local search capability, whereas the cognitive and social factors determine the impact of the personal best and the global best, respectively.

A larger inertia weight (w) facilitates a global search, whereas a smaller inertia weight tends to facilitate a local search to fine-tune the current search area [269]. A balance between global and local exploration abilities can be achieved by selecting the inertia weight suitably, which results in fewer iterations on an average to find the optimum. Besides, if cognitive factor (c_1) is greater than social factor (c_2), the particle tends to converge to the best position found by itself rather than the best position found by the population, and vice versa.

Since the PSO search process is nonlinear and complicated, the inertia weight and cognitive and social factors should be changed nonlinearly and dynamically to achieve better dynamics of balance between global and local search abilities and to handle various types of uncertainty or noise in the optimization problem of different domains.

Different existing variants of PSO can be found, which were used previously to determine the optimal values of the tuning parameters of PSO.

In time-varying acceleration coefficient PSO (TVAC-PSO) [293], the cognitive factor (c1) starts with a higher value than the social factor (c2) and linearly decreases while the social factor starts with a lower value and linearly increases. In time-varying inertia weight PSO (TVIW-PSO) [324], the inertia weight (w) decreases linearly from a relatively large value to a small value through the course of a PSO run.

Since the PSO search process is a nonlinear and complicated process, these approaches have a linear transition of search capability from global to local search. In dynamic optimization problems, these approaches do not truly reflect the actual search process, which is required to find the optimum.

In fuzzy adaptive PSO (FAPSO) [325], a fuzzy system was used to adjust the inertia weight of PSO. However, the fuzzy-based system cannot address all types of uncertainty. It can handle uncertainties due to imprecision, vagueness, and ambiguity, but it cannot address uncertainties due to incompleteness and ignorance, which can be observed with various dynamic optimization problems of different domains.

Therefore, a suitable knowledge representation schema and reasoning mechanism should be used for addressing different types of uncertainties existing with different dynamic optimization problems of various domains. An efficient way of addressing this issue can be the utilization of Belief Rule-Based Expert System (BRBES), which has a unique capability of dealing with the different types of uncertainties due to incompleteness, ignorance, vagueness, imprecision, randomness, and ambiguity.

BRBES is an improved version of expert systems that can represent uncertain knowledge by incorporating the belief structure. It uses Belief Rule Base (BRB) to represent uncertain knowledge, while Evidential Reasoning (ER) acts as an inference engine to handle both uncertain and heterogeneous data [386]. BRB has a powerful nonlinear modeling ability, which can be used to identify the behaviors of complex systems [398]. In general, there are two types of BRB, namely conjunctive and disjunctive. In conjunctive BRB, each rule is assumed as conjunctive in nature, while in disjunctive BRB, each rule is represented using a disjunctive assumption. However, due to the conjunctive nature [10] of the rule, conjunctive BRB suffers from the combinatorial explosion problem, whereas disjunctive BRB does not suffer from a similar problem and it requires less computational time to process data.

In this paper, a new Belief Rule-Based Adaptive Particle Swarm Optimization (BRBAPSO) is proposed where the tuning parameters are adjusted dynamically using BRBES, which considers uncertainties and ensures a balance between exploitation and exploration in the search space. Two variants of BRBAPSO, namely Conjunctive BRBAPSO and Disjunctive BRBAPSO, are introduced, and they are compared with time-varying inertia weight PSO (TVIW-PSO), time-varying acceleration coefficient PSO (TVAC-PSO), and Fuzzy Adaptive PSO (FAPSO) using the CEC 2013 real-parameter optimization benchmark functions to evaluate the effectiveness of the BRBAPSO algorithm.

The rest of the paper is organized as follows. Section 5.2 briefly discusses the basics of BRBES, while PSO is described in Section 5.3. Afterwards, the proposed BRBAPSO algorithm is discussed in Section 5.4. Section 12.6 provides an analysis of the results to demonstrate the effectiveness of the proposed method, and the paper is concluded in Section 5.6.

5.2 Belief Rule Based Expert System (BRBES)

A Belief Rule-Based Expert System (BRBES) consists of two main parts, namely a knowledge base and an inference engine. In BRBES, Belief Rule Base (BRB) is used to represent uncertain knowledge and to create the initial knowledge base, whereas Evidential Reasoning (ER) works as an inference engine by handling both heterogeneous and uncertain data [168]. The knowledge representation and inference mechanisms of BRBES are presented in this section.

5.2.1 Domain knowledge representation in BRBES

A Belief Rule Base (BRB) is an extended version of conventional IF-THEN rule base which can express more complicated non-linear causal connections under uncertainty. A belief rule comprises two main parts, namely antecedent and consequent. Each antecedent attribute is associated with referential values, while belief degrees are embedded with the referential values of the consequent attribute. BRB contains various learning or knowledge representation parameters, including attribute weight, rule weight, and belief degrees, which are used to capture uncertainty in data [195], [167]. A belief rule can be defined as follows:

$$R_k : \begin{cases} \text{IF } (A_1 \text{ is } V_1^k) \text{ AND / OR } (A_2 \text{ is } V_2^k) \text{ AND / OR } ...\text{AND / OR } (A_{T_k} \text{ is } V_{T_k}^k) \\ \text{THEN } C \text{ is } (C_1, \beta_{1k}), (C_2, \beta_{2k}), ..., (C_N, \beta_{Nk}) \end{cases}$$

$$\text{where } \beta_{jk} \geq 0, \sum_{j=1}^{N} \beta_{jk} \leq 1 \text{ with rule weight } \theta_k,$$

$$\text{and attribute weights } \delta_{k1}, \delta_{k2}, ..., \delta_{kT_k}, \ k \in 1, ..., L$$

In the above rule, $A_1, A_2, ..., A_{T_k}$ are the antecedent attributes of the kth rule. $V_i^k (i = 1, ..., T_k, k = 1, ..., L)$ is the referential value of the ith antecedent attribute, while C_j is the jth referential value of the consequent attribute. $\beta_{jk} (j = 1, ...N, k = 1, ...L)$ is the degree of belief to which the consequent reference value C_j is believed to be true. T_k is the total number of antecedent attributes used in the kth rule. L is the number of total belief rules and N is the number of all possible referential values of the consequent. If $\sum_{j=1}^{N} \beta_{jk} = 1$, the kth rule is said to be complete. If the summation of belief degrees is less than 1, the rule is considered as incomplete, which can happen because of ignorance or incompleteness. In the traditional IF-THEN rule, antecedents and the consequent attributes have a linear relationship while the relationship is non-linear in case of belief rule. Besides, data collected from interviews or surveys are naturally non-linear [361]. As a consequence, belief rules can be used in order to represent the data efficiently.

The logical connectives of the antecedent attributes in a belief rule can be either AND or OR, which represents the conjunctive or the disjunctive assumptions of the rule, respectively. Based on the logical connectivity of the Belief Rule Base, a BRBES can be named as conjunctive or disjunctive BRBES.

Under the conjunctive assumption, the total number of rules, L is calculated using the referential values, J_i of the antecedent attributes, A_i of a BRB, as shown in eq. (5.1).

$$L = \prod_{i=1}^{T_k} J_i \tag{5.1}$$

Under the disjunctive assumption, the total number of rules,L is equal to the number of referential values of the antecedent attributes, as shown in eq. (5.2). The disjunctive assumption requires that all attributes have the same number of referential values [380].

$$L = J_1 = J_2 = ... = J_i \tag{5.2}$$

5.2.2 BRB inference procedures

Evidential Reasoning (ER) can handle heterogeneous data as well as different types of uncertainties such as incompleteness, ignorance, imprecision, and vagueness [241], [240]. The inference procedures using the ER approach contain different steps, namely input transformation, rule activation weight calculation, belief update, and rule aggregation, which is shown in Fig. 5.1.

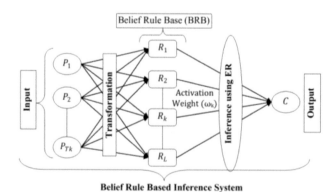

Belief Rule Based Inference System

Figure 5.1: Sequence of BRBES inference procedures.

5.2.2.1 Input transformation

During input transformation, the input data is distributed over the referential values of the antecedent attribute of a rule, as shown in eq. (5.3) [169].

$$H(v_i) = (V_{i,j}, \alpha_{i,j}), j = 1, ..., J_i, i = 1, ..., T_k \tag{5.3}$$

Here, the function H transforms the input value of the antecedent attribute to the belief degrees of its referential values, where V_{ij} is the jth referential value of the input, and α_{ij} is the belief degree to the referential value. These transformed values of the input data are known as matching degrees. The calculation is carried out using eqs. (5.4), (5.5), and (5.6).

$$\alpha_{i,j} = \frac{V_{i,j+1} - v_i}{V_{i,j+1} - V_{i,j}}, \ V_{i,j} \leq v_i \leq V_{i,j+1}, \ j = 1, 2, ..., J_i - 1 \quad (5.4)$$

$$\alpha_{i,j+1} = 1 - \alpha_{i,j}, \ V_{i,j} \leq v_i \leq V_{i,j+1}, \ j = 1, 2, ..., J_i - 1 \quad (5.5)$$

$$\alpha_{i,k} = 0, \ k = 1, 2, ..., J_i, \ k \neq j, j+1 \quad (5.6)$$

After assigning the matching degree, the rules are called packet antecedent, and they become active.

5.2.2.2 Rule activation weight calculation

In order to calculate rule activation weight, the first task is to combine the individual matching degrees of the antecedent attributes of a rule using a weighted multiplicative equation for the conjunctive assumption, as shown in eq. (5.7).

$$\alpha_{k_{conj}} = \prod_{i=1}^{T_k} (\alpha_i^k)^{\bar{\delta}_{ki}} \quad (5.7)$$

In case of the disjunctive assumption, the individual matching degrees are combined using eq. (5.8).

$$\alpha_{k_{disj}} = \sum_{i=1}^{T_k} (\alpha_i^k)^{\bar{\delta}_{ki}} \quad (5.8)$$

For both conjunctive and disjunctive assumptions,

$$\bar{\delta}_{ki} = \frac{\delta_{ki}}{max_{i=1,...T_k}\{\delta_{ki}\}}, \ 0 \leq \bar{\delta}_{ki} \leq 1$$

Here, T_k is the total number of antecedent attributes in the kth rule, δ_{ki} is the weight of each antecedent attribute V_i, and $\bar{\delta}_{ki}$ is the relative weight of V_i, which is calculated by dividing the weights of V_i by the maximum weight of all antecedent attributes.

Then, for the conjunctive assumption, the combined matching degree of each rule calculated by eq. (5.7) is utilized to determine the activation weight w_k for the kth rule, as shown in eq. (5.9) [166].

$$w_{k_{conj}} = \frac{\theta_k \alpha_{k_{conj}}}{\sum_{i=1}^{L}(\theta_i \alpha_{i_{conj}})} \tag{5.9}$$

Similarly, for the disjunctive assumption, the combined matching degree of each rule calculated by eq. (5.8) is used to measure the activation weight w_k for the kth rule, as shown in eq. (5.10).

$$w_{k_{disj}} = \frac{\theta_k \alpha_{k_{disj}}}{\sum_{i=1}^{L}(\theta_i \alpha_{i_{disj}})} \tag{5.10}$$

Here, θ_k represents the rule weight, while α_k represents the combined matching degree of the kth rule. The activation weight of a rule will be zero if that rule is not activated. After calculating the sum of the rule activation weight of a rule base, the result should be one [361].

5.2.2.3 Belief update

In some cases, if there is an absence of data for any antecedent attribute of a rule base because of ignorance, then the initial belief degrees embedded with each rule in the rule base need to be updated to address the uncertainty due to ignorance, which is shown in eq. (5.11).

$$\beta_{jk} = \bar{\beta}_{jk} \frac{\sum_{t=1}^{T_k}(\tau(t,k)\sum_{i=1}^{I_t}(\alpha_{ti}))}{\sum_{t=1}^{T_k}\tau(t,k)} \tag{5.11}$$

where, $\tau(t,k) = \begin{cases} 1 & \text{if the } t^{th} \text{ attribute is used in defining rule } R_k(k=1,...,T_k) \\ 0 & \text{otherwise} \end{cases}$

Here, $\bar{\beta}_{jk}$ is the original belief degree, while β_{jk} is the updated belief degree of the kth rule. α_{ti} represents the degree to which the input value belongs to an attribute.

5.2.2.4 Rule aggregation using Evidential Reasoning (ER)

All the packet antecedents of the rules need to be aggregated to calculate the output for the input data of the antecedent attributes using the Evidential Reasoning

(ER) algorithm. The aggregation of the rules can be done by using either analytical or recursive ER algorithms [165], [386]. However, the analytical approach is preferable instead of the recursive approach, since it is computationally more efficient [387]. The analytical ER computation can be performed using eq. (5.12) [372].

$$
\beta_j = \frac{\mu \times \left[\prod_{k=1}^{L} (w_k \beta_{jk} + 1 - w_k \sum_{j=1}^{N} \beta_{jk}) - \prod_{k=1}^{L} (1 - w_k \sum_{j=1}^{N} \beta_{jk}) \right]}{1 - \mu \times \left[\prod_{k=1}^{L} (1 - w_k) \right]} \tag{5.12}
$$

$$
\text{where, } \mu = \left[\sum_{j=1}^{N} \prod_{k=1}^{L} (w_k \beta_{jk} + 1 - w_k \sum_{j=1}^{N} \beta_{jk}) - (N-1) \times \prod_{k=1}^{L} (1 - w_k \sum_{j=1}^{N} \beta_{jk}) \right]^{-1}
$$

Here, w_k represents the activation weight of the kth rule, whereas β_j is the belief degree associated with one of the consequent reference values.

The uncertainty due to vagueness, imprecision, and ambiguity are addressed by eq. (5.12) during the process of rule aggregation [372]. Now the calculated output value against the input data will be in a fuzzy form. So this fuzzy value can be converted into a crisp or numerical value by using the utility score associated with each referential value of the consequent attribute to obtain the final result, which is shown in eq. (5.13).

$$
z_i = \sum_{j=1}^{N} \mu(O_j) \beta_j \tag{5.13}
$$

Here, z_i is the expected numerical value, while $\mu(O_j)$ is the utility score of each referential value.

5.3 Particle Swarm Optimization (PSO)

Particle Swarm Optimization (PSO) is a stochastic and population-based meta-heuristic algorithm, where a population or swarm contains a set of individuals. In PSO, each individual represents a possible solution to the problem, which is referred to as a particle. The particles move around in the search space to find the best solution. Each particle is characterized by three indicators, namely the position, the velocity, and the fitness value. The velocity and the position of each particle are initialized randomly within the corresponding ranges. Then the values of the fitness function for each particle are evaluated for each particle, and they are set as their own local optimal solution. The particle that has the minimum value for the fitness function is selected as the global solution. Next, the velocity and the position is updated by using eqs. (5.14) and (5.15).

$$v_{id} = w * v_{id} + c_1 * rand_1() * (p_{id} - x_{id}) + c_2 * rand_2() * (p_{gd} - x_{id}) \quad (5.14)$$

$$x_{id} = x_{id} + v_{id} \quad (5.15)$$

In the above equations, v_{id} and x_{id} are the velocity and the position of the dth dimension of the particle $i(i = 1, 2, ..., SwarmSize)$. w is the inertia weight, c_1 is the cognitive factor, and c_2 is the social factor. $rand_1()$ and $rand_2()$ are two random numbers in the range $[0, 1]$. $p_i = (p_{i1}, p_{i2}, ..., p_{id})$ is the optimal position of the ith particle, and $p_g = (p_{g1}, p_{g2}, ..., p_{gd})$ is the global optimal position of all particles.

The velocity is restricted in the range of the velocity boundary $[V_{min}, V_{max}]$ and the position is also restricted in the range of the search space $[X_{min}, X_{max}]$ as shown in eqs. (5.16), (5.17):

$$v_{id} = \begin{cases} V_{max}, & \text{if } v_{id} > V_{max} \\ V_{min}, & \text{if } v_{id} < V_{min} \end{cases} \quad (5.16)$$

where, $V_{max} = 0.1 * (X_{max} - X_{min}); V_{min} = -V_{max}$

$$x_{id} = \begin{cases} X_{max}, & \text{if } x_{id} > X_{max} \\ X_{min}, & \text{if } x_{id} < X_{min} \end{cases} \quad (5.17)$$

where, X_{max} and X_{min} are the upper and lower bounds of the seach space.

Afterwards, the fitness function for each particle is evaluated again. If the value of the fitness function for any particle is better than its local optimal solution, then the solution is updated, and the particle that has the minimum value for the objective function is selected as the global solution. The search procedure is stopped if the current iteration number reaches the predetermined maximum iteration number. The flowchart of the PSO algorithm is illustrated in Fig. 5.2.

5.4 Belief Rule Based Adaptive Particle Swarm Optimization (BRBAPSO)

The three parameters, namely inertia weight (w), cognitive factor (c_1), and social factor(c_2) have a great impact on the performance of PSO. These parameters can be adapted dynamically to improve the performance of PSO. It is suggested that changing the values of w, c_1, and c_2 dynamically during each iteration of PSO

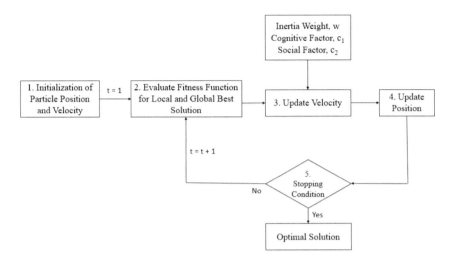

Figure 5.2: The flowchart of PSO algorithm.

can produce better results. Therefore, a BRBES based PSO parameter adaptation algorithm is proposed, named Belief Rule-Based Adaptive Particle Swarm Optimization (BRBAPSO), which can change the value of these parameters dynamically during each iteration of the algorithm.

In BRBAPSO, first, the diversity of swarm is calculated as the average of the Euclidean measure of distance between each particle and the jth dimension over all particles, as shown in eq. (5.18).

$$\text{DiversitySwarm, } d_s = \frac{1}{N_s} \sum_{i=1}^{N_s} \sqrt{\sum_{j=1}^{N_x} (x_{ij}(t) - \bar{x}_j(t))^2} \qquad (5.18)$$

$$\text{where, } \bar{x}_j(t) = \frac{\sum_{i=1}^{N_s} x_{ij}(t)}{N_s}$$

Then the diversity of velocity is calculated as the average of the Euclidean measure of velocity between each particle and the jth dimension over all particles, as shown in eq. (5.19).

$$\text{DiversityVelocity, } d_v = \frac{1}{N_s} \sum_{i=1}^{N_s} \sqrt{\sum_{j=1}^{N_x} (v_{ij}(t) - \bar{v}_j(t))^2} \qquad (5.19)$$

$$\text{where, } \bar{v}_j(t) = \frac{\sum_{i=1}^{N_s} v_{ij}(t)}{N_s}$$

Afterward, the diversity of swarm and the diversity of velocity is normalized in the range $[0, 1]$, as shown in eqs. (5.20) and (5.21).

Normalized Diversity of Swarm, $nd_s =$

$$\begin{cases} 0, & \text{if Min DiverSwarm = Max DiverSwarm} \\ \frac{\text{DiversitySwarm} - \text{Min DiverSwarm}}{\text{Max DiverSwarm} - \text{Min DiverSwarm}}, & \text{if Min DiverSwarm} \neq \text{Max DiverSwarm} \end{cases}$$

(5.20)

Normalized Diversity of Velocity, $nd_v =$

$$\begin{cases} 0, & \text{if Min DiverVel = Max DiverVel} \\ \frac{\text{DiversityVelocity} - \text{Min DiverVel}}{\text{Max DiverVel} - \text{Min DiverVel}}, & \text{if Min DiverVel} \neq \text{Max DiverVel} \end{cases}$$

(5.21)

Here, Min DiverSwarm and Max DiverSwarm are the minimum and maximum diversity of swarm, while Min DiverVel and Max DiverVel are the minimum and maximum diversity of velocity respectively.

Generally, when the best fitness is low at the end of the run in the optimization of a minimum function, low inertia weight (w) and high learning factors (c_1 and c_2) are often preferred [268]. On the contrary, when the best fitness stays at one value for a long time, the number of unchanged best fitness generations is large, that is, the system is stuck at a local minimum. So, the system should focus on exploiting rather than exploring. Therefore, the inertia weight (w) should be increased, and learning factors (c_1 and c_2) should be decreased [7].

According to this knowledge, the inertia weight (w) and learning factors ($c1$ and $c2$) are adjusted in BRBAPSO using BRBES. In BRBAPSO, the values of w, c_1, and c_2 are initially taken as 1, 2, and 2 respectively. Then after the first iteration, normalized diversity of swarm and normalized diversity of velocity are supplied as inputs to BRBES. Afterwards, based on the Belief Rule Base and the use of Evidential Reasoning approach, new values of w, c_1, and c_2 are produced by BRBES as outputs after the first iteration. This process continues until the current iteration number reaches the predetermined maximum. The system diagram of BRBAPSO is depicted in Fig. 5.3.

In this paper, two variants of BRBAPSO are developed. They are Conjunctive and Disjunctive BRBAPSO. The Belief Rule Base (BRB) used for conjunctive BRBAPSO and disjunctive BRBAPSO is shown in Table 5.1 and Table 5.2, respectively.

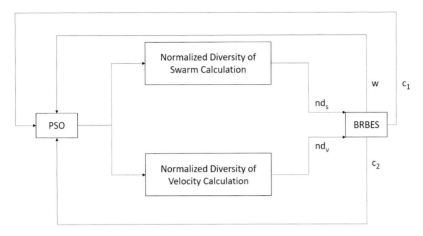

Figure 5.3: Belief Rule Based Adaptive Particle Swarm Optimization (BRBAPSO).

Table 5.1: Belief rule base for conjunctive BRBAPSO.

Rule ID	Rule Weight	IF		THEN ($w/c_1/c_2$)		
		nd_s	nd_v	High	Medium	Low
1	1	High	High	1	0	0
1	1	High	Medium	0.5	0.5	0
1	1	High	Low	0	1	0
1	1	Medium	High	0.5	0.5	0
1	1	Medium	Medium	0	1	0
1	1	Medium	Low	0	0.5	0.5
1	1	Low	High	0	1	0
1	1	Low	Medium	0	0.5	0.5
1	1	Low	Low	0	0	1

Table 5.2: Belief rule base for disjunctive BRBAPSO.

Rule ID	Rule Weight	IF		THEN ($w/c_1/c_2$)		
		nd_s	nd_v	High	Medium	Low
1	1	High	High	1	0	0
1	1	Medium	Medium	0	1	0
1	1	Low	Low	0	0	1

To achieve optimal exploration and exploitation of the search space, the iteration of PSO is divided into eight stages and four BRBES are employed in both conjunctive and disjunctive BRBAPSO.

In the first stage, $BRBES_1$ is used for predicting the value of w, c_1, and c_2, respectively. Similarly, in the second stage, $BRBES_2$ is utilized to predict the new value of w, c_1, and c_2. The referential values and utility values of both antecedent and consequent attributes of $BRBES_1$ and $BRBES_2$ are shown in Table 5.3 and Table 5.4. These two stages will help particles to explore individually and achieve their own best historical positions, rather than crowd around the current best particle that may be associated with a local optimum.

Table 5.3: Details of $BRBES_1$ (H: High, M: Medium, L: Low).

| | Antecedent Attributes | | | | | |
| | nd_s | | | nd_v | | |
Referential Values	H	M	L	H	M	L
Utility Values	0.80	0.40	0.01	0.80	0.40	0.01

| | Consequent Attributes | | | | | | | | |
| | w | | | c_1 | | | c_2 | | |
Referential Values	H	M	L	H	M	L	H	M	L
Utility Values	1.0	0.93	0.85	2.1	2.2	2.3	2.0	1.9	1.8

Table 5.4: Details of $BRBES_2$ (H: High, M: Medium, L: Low).

| | Antecedent Attributes | | | | | |
| | nd_s | | | nd_v | | |
Referential Values	H	M	L	H	M	L
Utility Values	0.80	0.40	0.01	0.80	0.40	0.01

| | Consequent Attributes | | | | | | | | |
| | w | | | c_1 | | | c_2 | | |
Referential Values	H	M	L	H	M	L	H	M	L
Utility Values	0.85	0.78	0.70	2.3	2.4	2.5	1.7	1.6	1.5

Likewise, in the third stage, $BRBES_3$ is used for predicting the value of w, c_1, and c_2, and in the fourth stage, $BRBES_4$ is utilized to predict the new value of w, c_1, and c_2, respectively. The referential values and utility values of both antecedent and consequent attributes of $BRBES_3$ and $BRBES_4$ are shown in Table 5.5 and Table 5.6. These two stages will help to emphasize the search and exploitation around the personal best of each particle and avoid the deception of a local optimum.

Afterward, in the fifth stage, $BRBES_1$, that is used in the first stage, is utilized again to predict the value of w, c_1, and c_2, respectively. In the sixth stage, $BRBES_2$, that is utilized in the second stage, is used again to predict the new value of w, c_1, and c_2. These two stages will help to guide other particles to the probable globally optimal region and avoid premature convergence.

Table 5.5: Details of $BRBES_3$ (H: High, M: Medium, L: Low).

	Antecedent Attributes					
	nd_s			nd_v		
Referential Values	H	M	L	H	M	L
Utility Values	0.80	0.40	0.01	0.80	0.40	0.01

	Consequent Attributes								
	w			c_1			c_2		
Referential Values	H	M	L	H	M	L	H	M	L
Utility Values	0.70	0.63	0.55	2.0	1.9	1.8	2.1	2.2	2.3

Table 5.6: Details of $BRBES_4$ (H: High, M: Medium, L: Low).

	Antecedent Attributes					
	nd_s			nd_v		
Referential Values	H	M	L	H	M	L
Utility Values	0.80	0.40	0.01	0.80	0.40	0.01

	Consequent Attributes								
	w			c_1			c_2		
Referential Values	H	M	L	H	M	L	H	M	L
Utility Values	0.55	0.48	0.40	1.7	1.6	1.5	2.3	2.4	2.5

In the seventh stage, $BRBES_3$, that is used in the third stage, is utilized again to predict the value of w, c_1, and c_2, and in the eighth stage, $BRBES_4$, that is utilized in the fourth stage, is used again to predict the new value of w, c_1, and c_2, respectively. These two stages will help the globally best particle to jump out of local optimum toward a better optimum and help other particles to fly to this new region as fast as possible.

The BRBAPSO algorithm is presented as pseudocode in Algorithm 6.

Algorithm 6 BRBAPSO Algorithm

1: **procedure** BRBAPSO(D, NP, Max_It)   ▷ D: dimension, NP: population size, Max_It: maximum iteration number
2:    **for** *each particle* $i \in 1,...,NP$ **do**
3:        Randomly initialize x_i
4:        Initialize v_i to zero
5:        Evaluate the fitness value of each particle and set all initial positions as $P_{Best_{x_i}}$
6:    **end for**
7:    Select the G_{Best} particle in the swarm, which has the minimum fitness value
8:    $w = 1$
9:    $c_1 = 2$
10:    $c_2 = 2$
11:    $it = 1$ ▷ it: current iteration number
12:    **while** $Max\_It \geq it$ or *(!stop criterion)* **do**
13:        **if** $it > 1$ **then**
14:            Calculate nd_s and nd_v using eqs. (5.20) and (5.21)
15:            **if** $it \leq Max\_It * (1/8)$ **then**
16:                $[newW, newC1, newC2] = BRBES_1(nd_s, nd_v)$
17:                $w = newW$
18:                $c_1 = newC1$
19:                $c_2 = newC2$
20:            **end if**
21:            **if** $it > Max\_It * (1/8)$ AND $it \leq Max\_It * (2/8)$ **then**
22:                $[newW, newC1, newC2] = BRBES_2(nd_s, nd_v)$
23:                $w = newW$
24:                $c_1 = newC1$
25:                $c_2 = newC2$
26:            **end if**
27:            **if** $it > Max\_It * (2/8)$ AND $it \leq Max\_It * (3/8)$ **then**
28:                $[newW, newC1, newC2] = BRBES_3(nd_s, nd_v)$
29:                $w = newW$
30:                $c_1 = newC1$
31:                $c_2 = newC2$
32:            **end if**
33:            **if** $it > Max\_It * (3/8)$ AND $it \leq Max\_It * (4/8)$ **then**
34:                $[newW, newC1, newC2] = BRBES_4(nd_s, nd_v)$
35:                $w = newW$
36:                $c_1 = newC1$
37:                $c_2 = newC2$
38:            **end if**
39:            **if** $it > round(Max\_It * (4/8)$ AND $it \leq Max\_It * (5/8)$ **then**
40:                $[newW, newC1, newC2] = BRBES_1(nd_s, nd_v)$
41:                $w = newW$
42:                $c_1 = newC1$
43:                $c_2 = newC2$
44:            **end if**
45:            **if** $it > Max\_It * (5/8)$ AND $it \leq Max\_It * (6/8)$ **then**
46:                $[newW, newC1, newC2] = BRBES_2(nd_s, nd_v)$
47:                $w = newW$

48:	$c_1 = newC1$
49:	$c_2 = newC2$
50:	**end if**
51:	**if** $it > Max\_It * (6/8)$ AND $it \leq Max\_It * (7/8)$ **then**
52:	$[newW, newC1, newC2] = BRBES_3(nd_s, nd_v)$
53:	$w = newW$
54:	$c_1 = newC1$
55:	$c_2 = newC2$
56:	**end if**
57:	**if** $it > Max\_It * (7/8)$ **then**
58:	$[newW, newC1, newC2] = BRBES_4(nd_s, nd_v)$
59:	$w = newW$
60:	$c_1 = newC1$
61:	$c_2 = newC2$
62:	**end if**
63:	**end if**
64:	**for** *each particle* $i \in 1,...,NP$ **do**
65:	Calculate the velocity (v_i) of each particle using eq. (5.14) and check the velocity in the range of the velocity boundary using eq. (5.16)
66:	Calculate the position (x_i) of each particle using eq. (5.15) and check the position in the range of the search space using eq. (5.17)
67:	Evaluate the fitness value of each particle and if the fitness value of any particle is better than its personal best as $P_{Best_{x_i}}$, current fitness value is set as the new particle best of that particle
68:	**end for**
69:	Select the G_{Best} particle in the swarm, which has the minimum fitness value
70:	$it = it + 1$
71:	**end while**
72:	**end procedure**

Therefore, the proposed BRBAPSO provides a solution for dealing with the uncertainty in objective functions by incorporating BRBES with PSO. Furthermore, it also helps to find optimal exploration and exploitation of the search space, which leads to finding the optimal solution.

5.5   Results and analysis

This section presents an empirical evaluation of BRBAPSO, which is evaluated on the CEC2013 benchmark problem set [220]. Two variants of BRBAPSO, namely conjunctive and disjunctive BRBAPSO are both compared with time-varying inertia weight PSO (TVIW-PSO), and acceleration coefficient PSO (TVAC-PSO), and Fuzzy Adaptive PSO (FAPSO).

The CEC2013 benchmark set consists of 28 test functions, where five functions $(F_1 \sim F_5)$ are unimodal, fifteen functions $(F_6 \sim F_{20})$ are multimodal, and eight functions $(F_{21} \sim F_{28})$ are composite functions, which combine multiple test problems into a complex landscape. For all of the problems, the search space is $[-100, 100]^D$.

The evaluation is performed following the guidelines of the CEC2013 benchmark competition [220]. When the difference between the values of the best solution found and the optimal solution was 10^{-8} or smaller, the error value was treated as 0. For all of the problems the number of dimensions $D = 10, 30$ and the maximum number of objective function calls per run was $D \times 10,000$ (i.e., $100,000$ and $300,000$, respectively). The number of runs per problem was 51, and the average performance of these runs was evaluated. Statistical significance testing was done using the Wilcoxon signed ranked test (significant threshold: $p < 0.05$).

The results of Conjunctive BRBAPSO, TVIW-PSO, TVAC-PSO, and FAPSO on D = 10 and D = 30 dimensions are shown in Table 5.7 and Table 5.8, where each column shows the mean of the error (difference) values between the best fitness values found in each run and the true optimal value.

The aggregate results of statistical testing $(+,-,\approx)$ comparing each algorithm vs. Conjunctive BRBAPSO on 28 functions is shown in Table 5.9. The symbols $+, -, \approx$ indicate that a given algorithm performed significantly better $(+)$, significantly worse $(-)$, or not significantly better or worse $(\approx)$ compared to Conjunctive BRBAPSO using the Wilcoxon signed ranked test (significant threshold: $p < 0.05$). The maximum number of objective function evaluations is $D \times 10,000$ (i.e., $100,000$ and $300,000$, respectively). All results are based on 51 runs.

According to the Wilcoxon test on the column $D = 10$ of Table 5.9, which summarizes the experimental results on $D = 10$, TVIW-PSO performed (significantly, $p < 0.05$) better than Conjunctive BRBAPSO on 7 functions and worse than Conjunctive BRBAPSO on 21 functions. TVAC-PSO performed (significantly, $p < 0.05$) better than Conjunctive BRBAPSO on 1 functions and worse than Conjunctive BRBAPSO on 27 functions. FAPSO performed (significantly, $p < 0.05$) better than Conjunctive BRBAPSO on 6 functions and worse than Conjunctive BRBAPSO on 21 functions and there is no significant difference for 1 functions.

Similarly, on the column $D = 30$ of Table 5.9, which summarizes the experimental results on $D = 30$, TVIW-PSO, TVAC-PSO and FAPSO performed (significantly, $p < 0.05$) better than Conjunctive BRBAPSO on 1 functions and worse than Conjunctive BRBAPSO on 27 functions.

Table 5.7: The mean of the error values of Conjunctive BRBAPSO, TVIW-PSO, TVAC-PSO, FAPSO, and Disjunctive BRBAPSO on the CEC2013 benchmark functions for $D = 10$ dimensions (Conjunctive BRBAPSO: CBRBAPSO, Disjunctive BRBAPSO: DBRBAPSO).

F	CBRBAPSO	TVIW-PSO	TVAC-PSO	FAPSO	DBRBAPSO
F_1	0.00e+00	1.60e+03	3.49e+02	1.26e+02	0.00e+00
F_2	1.17e+05	3.98e+06	1.46e+06	4.22e+05	1.29e+05
F_3	9.98e+06	2.55e+09	2.46e+09	6.83e+08	1.77e+07
F_4	3.24e+02	9.76e+03	5.17e+03	4.29e+03	4.52e+02
F_5	3.92e+00	2.83e+02	1.39e+02	4.05e+01	0.00e+00
F_6	1.20e+01	4.90e+01	4.10e+01	3.59e+01	1.33e+01
F_7	1.01e+01	7.33e+01	4.99e+01	6.05e+01	9.54e+00
F_8	2.03e+01	2.03e+01	2.03e+01	2.03e+01	2.03e+01
F_9	3.16e+00	7.13e+00	4.88e+00	4.99e+00	3.09e+00
F_{10}	2.86e+00	1.71e+02	7.10e+01	3.12e+01	1.68e+00
F_{11}	2.58e+00	7.51e+01	1.17e+01	6.81e+00	2.54e+00
F_{12}	1.25e+01	7.42e+01	3.00e+01	2.09e+01	1.28e+01
F_{13}	1.79e+01	7.34e+01	3.75e+01	3.02e+01	1.75e+01
F_{14}	3.58e+02	1.57e+03	2.96e+02	2.93e+02	3.51e+02
F_{15}	5.49e+02	1.47e+03	9.09e+02	8.93e+02	5.09e+02
F_{16}	8.51e-01	1.13e+00	7.63e-01	7.75e-01	8.82e-01
F_{17}	1.42e+01	1.29e+02	1.29e+01	1.30e+01	1.38e+01
F_{18}	3.22e+01	1.34e+02	2.88e+01	2.77e+01	3.22e+01
F_{19}	6.09e-01	3.01e+01	9.59e+01	1.08e+01	6.04e-01
F_{20}	2.78e+00	3.59e+00	3.20e+00	3.23e+00	2.78e+00
F_{21}	3.92e+02	5.01e+02	3.64e+02	3.73e+02	4.00e+02
F_{22}	4.57e+02	1.35e+03	4.54e+02	4.36e+02	4.13e+02
F_{23}	7.98e+02	1.43e+03	1.00e+03	1.06e+03	8.33e+02
F_{24}	2.11e+02	2.20e+02	2.22e+02	2.22e+02	2.10e+02
F_{25}	2.12e+02	2.20e+02	2.20e+02	2.19e+02	2.09e+02
F_{26}	1.99e+02	1.95e+02	2.27e+02	2.23e+02	2.02e+02
F_{27}	3.87e+02	5.83e+02	5.84e+02	5.63e+02	3.84e+02
F_{28}	3.55e+02	8.36e+02	3.51e+02	3.99e+02	3.34e+02

The aggregate results of statistical testing $(+, -, \approx)$ comparing each algorithm vs. Disjunctive BRBAPSO on 28 functions is shown in Table 5.10.

According to the Wilcoxon test on the column $D = 10$ of Table 5.10, TVIW-PSO performed (significantly, $p < 0.05$) better than Disjunctive BRBAPSO on 5 functions and worse than Disjunctive BRBAPSO on 23 functions. TVAC-PSO performed (significantly, $p < 0.05$) better than Disjunctive BRBAPSO on 1 functions and worse than Disjunctive BRBAPSO on 27 functions. FAPSO performed (significantly, $p < 0.05$) better than Disjunctive BRBAPSO on 5 functions and worse than Disjunctive BRBAPSO on 22 functions and there is no significant difference for 1 functions.

Table 5.8: The mean of the error values of Conjunctive BRBAPSO, TVIW-PSO, TVAC-PSO, FAPSO, and Disjunctive BRBAPSO on the CEC2013 benchmark functions for $D = 30$ dimensions (Conjunctive BRBAPSO: CBRBAPSO, Disjunctive BRBAPSO: DBRBAPSO).

F	CBRBAPSO	TVIW-PSO	TVAC-PSO	FAPSO	DBRBAPSO
F_1	5.11e+02	2.09e+04	1.28e+04	5.28e+03	4.01e+02
F_2	4.36e+06	1.10e+08	9.04e+07	2.03e+07	4.42e+06
F_3	5.66e+09	6.25e+10	9.28e+10	5.36e+10	5.23e+09
F_4	7.76e+02	4.40e+04	3.10e+04	1.56e+04	7.97e+02
F_5	2.23e+02	5.01e+03	4.27e+03	1.95e+03	5.26e+02
F_6	8.17e+01	9.44e+02	7.82e+02	2.96e+02	7.61e+01
F_1	6.04e+01	2.22e+02	2.21e+02	2.00e+02	5.29e+01
F_1	2.09e+01	2.09e+01	2.09e+01	2.09e+01	2.09e+01
F_9	1.80e+01	3.65e+01	2.56e+01	2.67e+01	1.83e+01
F_{10}	1.26e+02	2.31e+03	1.68e+03	9.43e+02	1.72e+02
F_{11}	3.41e+01	4.64e+02	2.63e+02	1.35e+02	3.54e+01
F_{12}	7.17e+01	4.69e+02	2.52e+02	1.78e+02	6.83e+01
F_{13}	1.36e+02	4.66e+02	3.01e+02	2.46e+02	1.34e+02
F_{14}	1.32e+03	6.03e+03	3.83e+03	2.43e+03	1.39e+03
F_{15}	3.61e+03	7.31e+03	5.40e+03	4.67e+03	3.46e+03
F_{16}	2.02e+00	2.44e+00	2.02e+00	2.02e+00	2.03e+00
F_{17}	6.03e+01	1.05e+03	2.41e+02	1.00e+02	6.12e+01
F_{18}	2.15e+02	1.03e+03	3.47e+02	2.51e+02	2.14e+02
F_{19}	3.25e+01	1.97e+04	3.41e+04	2.19e+03	1.62e+02
F_{20}	1.23e+01	1.30e+01	1.23e+01	1.26e+01	1.21e+01
F_{21}	3.41e+02	2.59e+03	1.47e+03	6.39e+02	2.96e+02
F_{22}	1.61e+03	6.79e+03	3.88e+03	2.55e+03	1.62e+03
F_{23}	4.26e+03	7.35e+03	5.58e+03	5.07e+03	4.22e+03
F_{24}	2.66e+02	2.91e+02	2.91e+02	2.92e+02	2.64e+02
F_{25}	2.86e+02	2.95e+02	3.01e+02	3.03e+02	2.84e+02
F_{26}	3.02e+02	2.87e+02	3.66e+02	3.54e+02	3.15e+02
F_{27}	8.20e+02	1.15e+03	1.06e+03	1.10e+03	8.24e+02
F_{28}	7.48e+02	3.20e+03	2.94e+03	2.33e+03	9.36e+02

Similarly, on the column $D = 30$ of Table 5.10, TVIW-PSO, TVAC-PSO and FAPSO performed (significantly, $p < 0.05$) better than Disjunctive BRBAPSO on 1 functions and worse than Disjunctive BRBAPSO on 27 functions.

As shown in Table 5.9 and Table 5.10, counting the number of +, −, and ≈ results, both variants of BRBAPSO, namely Conjunctive and Disjunctive BRBAPSO, clearly have the best overall performance on these 28 functions for all $D \in \{10, 30\}$. These results show that BRBAPSO, the proposed methods, significantly outperform previous TVIW-PSO, TVAC-PSO, and FAPSO algorithms overall on the CEC2013 benchmarks.

Table 5.9: The aggregate results of statistical testing $(+, -, \approx)$ comparing Conjunctive BR-BAPSO with TVIW-PSO, TVAC-PSO, and FAPSO algorithms on the CEC2013 benchmark functions ($D = 10, 30$ dimensions).

vs. Conjunctive BRBAPSO		$D = 10$	$D = 30$
	+ (better)	7	1
TVIW-PSO	− (worse)	21	27
	≈ (no sig.)	0	0
	+ (better)	1	1
TVAC-PSO	− (worse)	27	27
	≈ (no sig.)	0	0
	+ (better)	6	1
FAPSO	− (worse)	21	27
	≈ (no sig.)	1	0

Table 5.10: The aggregate results of statistical testing $(+, -, \approx)$ comparing Disjunctive BR-BAPSO with TVIW-PSO, TVAC-PSO, and FAPSO algorithms on the CEC2013 benchmark functions ($D = 10, 30$ dimensions).

vs. Disjunctive BRBAPSO		$D = 10$	$D = 30$
	+ (better)	5	1
TVIW-PSO	− (worse)	23	27
	≈ (no sig.)	0	0
	+ (better)	1	1
TVAC-PSO	− (worse)	27	27
	≈ (no sig.)	0	0
	+ (better)	5	1
FAPSO	− (worse)	22	27
	≈ (no sig.)	1	0

5.6 Conclusion

In this study, a new Belief Rule-Based Adaptive Particle Swarm Optimization (BRBAPSO) has been proposed, which exploits a BRBES to dynamically adjust the values of inertia weight, cognitive factor, and the social factor of PSO and ensures a balanced exploration and exploitation of search space. Two variants of BRBPSO, namely conjunctive and disjunctive BRBAPSO, were compared to existing time-varying inertia weight PSO (TVIW-PSO), time-varying acceleration coefficient PSO (TVAC-PSO), and Fuzzy Adaptive PSO (FAPSO) on the CEC 2013 real-parameter optimization benchmark functions. The results show that the proposed variants of BRBAPSO outperform the existing methods for different functions.

Chapter 6

NoSQL Environments and Big Data Analytics for Time Series

*Ciprian-Octavian Truica** and *Elena-Simona Apostol*

6.1 Introduction

In recent years, NoSQL technologies and Big Data analytics have gained increased attention from professionals and the research community. The ability of these solutions to store, process, and analyze large datasets also captured the interest of researchers and practitioners working with Time Series data. Thus two research directions have emerged: Time Series Data Management and Time Series Analysis. Time Series data management objectives are: (i) optimal Time Series data processing and storing, and (ii) efficient Time Series data querying and retrieving. The main goals of Time Series Analysis are: (i) identifying the nature of the phenomenon represented by the sequence of observations, (ii) predicting future values of the Time Series variable, (iii) outlier detection of extreme values that deviate from other observations, and (iv) change point detection for understanding trends and seasonality of the Time Series. In this chapter, we present an overview of Time Series Analysis and NoSQL Time Series databases to enable the use of Big Data machine learning techniques.

Computer Science and Engineering Department, Faculty of Automatic Control and Computers, University Politehnica of Bucharest.

* Corresponding author: ciprian.truica@cs.pub.ro

With the current development in the Internet of Things, the need to analyze sequential, unstructured, schema-less data has arisen. The recent advances have enabled rethinking of how data collecting, preprocessing, management, storing, and analysis is done. Collecting large volumes of data at high velocity requires management systems that provide high availability and high throughput. This can be achieved with dedicated Time Series database systems that optimize data storing and management by changing the architectural model. Retrieving large volumes of Time Series data promptly requires new data models for querying, transforming, and aggregating data using languages similar to SQL optimized to work with concepts such as multi-dimensional sequences. Time Series analysis can be achieved using dedicated frameworks for data processing, Machine Learning, and Deep Learning.

This chapter is structured as follows. Section 6.2 presents an introduction to Time Series and analysis models and techniques for forecasting, outlier detection, and change point detection. Section 6.3 introduces concepts from Database Management Systems with a focus on NoSQL databases. Section 6.4 presents some of the most relevant NoSQL Time Series databases and provides a comparison using key architectural and data modeling traits. Section 6.5 discusses how Edge, Fog, Cloud Computing, and IoT enables Big Data Analytics on Time Series.

6.2 Time series analysis

A Time Series is a set of observations recorded at a specific point in time. A Time series can be viewed as discrete, with each observation measured at a fixed point in time, or in continuous intervals, the values of measurements being recorded continuously in each window of time [58]. In this chapter, we will define a Time Series and then discuss some of the main directions taken for Time Series data analysis.

6.2.1 Understanding time series

Time series are used to develop statistical mathematical models that can describe time sequence data points [326]. A Time Series can be described as a sequence of random variables $X = \{x_1, x_2, ..., x_T\} = \{x_t | t \in \overline{1, T}\}$, where t denotes the instant of time represented by the point when a value is measured. The cardinality of X is T, that is, $T = ||X||$. Each time point can be continuous or discrete. Moreover, the sequence X is a stochastic process meaning that it can be defined as a family of random variables.

A Time Series can be univariate or multivariate. A univariate Time Series is a series with a single time dependent variable. A multivariate Time Series has at least two time dependent variables, $M \geq 2$ where "*M*" represents their number. Each of these variables depends on both their past values and other variables.

The classical additive decomposition of a Time Series is given by eq. (6.1), while the multiplicative decomposition is presented in eq. (6.2). Both models incorporate the seasonal trend, and the noise (residual) component.

$$x_t = m_t + s_t + y_t \qquad (6.1)$$

$$x_t = m_t \cdot s_t \cdot x_t \qquad (6.2)$$

The trend component is a slowly changing function (m_t) and represents variations of low frequency in a Time Series, the high and medium frequency fluctuations having been filtered out. The trend can be determined by the moving averages or spectral smoothing methods.

The seasonal component is a function (s_t) that represents fluctuations which are more or less stable after a known period h, also known as lag. These variations in the Time Series are considered normal.

The seasonal and trend components indicate the autocorrelation of a Time Series X. The autocorrelation property determines if a Time Series is linearly related to a lagged version of itself. Thus, given the autocovariance function at a lag h as $\gamma_X(h) = Cov(X_{t+h}, X_t)$, where $Cov(\cdot, \cdot)$ is the covariance function, then the autocorrelation is $\rho_X(h) = \frac{\gamma_X(h)}{\gamma_X(0)}$.

The noise component $Y = \{y_1, y_2, ..., y_T\} = \{y_t | t \in \overline{1, T}\}$ or residual sequences are used to check if an analysis model has correctly determined the information in the Time Series data and can help to predict future values [59]. A Time Series X is stationary if its properties are similar to a h time lag shifted Time Series. Thereby, X is stationary if its mean ($\mu_X(t)$) is independent of t and its covariance ($\gamma_X(t+h, t)$) is independent of t for each h.

Usually, the trend and seasonality components are eliminated to get stationary residuals.

6.2.2 Time series forecasting

Forecasting comprises different statistical and machine learning methods for predicting the future as accurately as possible. These models rely on historical data and knowledge of any future event that might impact the forecasts [170]. The statistical approaches use regression models, while the machine learning approaches use algorithms such as Support Vector Machines and Neural Networks.

6.2.2.1 Regression models

In this section, we present multiple regression models. Regression models are used for predicting the value of a given continuous variable based on the values of other variables. The model assumes a linear or nonlinear model of dependency. Regression models are used to understand the relationships between independent variables and dependent variables. Furthermore, the model also tries to infer causal relationships between independent variables and dependent variables.

Linear Regression

Linear Regression is a basic statistical model used for prediction. Given a univariate Time Series $X = \{x_t | t \in \overline{1,T}\}$, the regression model tries to find a linear equation that creates the line of best fit between the forecast variable y_t and a predictor x_t using eq. (6.3), where β_0 represents the predicted value of y_t when $x_t = 0$, β_1 represents the average predicted change in y_t resulting from a one unit increase in x_t, and ε is a random error that denotes a deviation from the underlying straight line model. The model tries to estimate β_0 and β_1, that is, $\hat{\beta}_0$, respectively $\hat{\beta}_0$, and determine the best fitting line $\hat{y}_t$ (eq. (6.4)) by minimizing the prediction error $e_t = y_t - \hat{y}_t$. The estimators are computed by minimizing the sum of the squared prediction errors (eq. (6.5)).

$$y_t = \beta_0 + \beta_1 \cdot x_t + \varepsilon_t \tag{6.3}$$

$$\hat{y}_t = \hat{\beta}_0 + \hat{\beta}_1 \cdot x_t \tag{6.4}$$

$$\min_{\hat{\beta}_0, \hat{\beta}_1} \sum_{t=1}^{T} e_t^2 = \min_{\hat{\beta}_0, \hat{\beta}_1} \sum_{t=1}^{T} (y_t - \hat{y}_t)^2 = \min_{\hat{\beta}_0, \hat{\beta}_1} \sum_{t=1}^{T} (y_t - \hat{\beta}_0 - \hat{\beta}_1)^2 \tag{6.5}$$

Multiple Linear Regression

In the case when the Time Series X is multivariate, then it becomes $X = \{x_{t,k} | t \in \overline{1,T} \wedge k \in \overline{1,M}\}$, with M denoting number of variables for each time point. Equation (6.6) presents the Multiple Linear Regression model.

$$y_t = \beta_0 + \beta_1 \cdot x_{t,1} + \beta_2 \cdot x_{t,2} + ... + \beta_M \cdot x_{t,M} + \varepsilon_t \tag{6.6}$$

The Multiple Linear Regression model can be rewritten using matrix notation (eq. (6.7)). The vector of estimated coefficients $\hat{\beta}$ is determined using the least square estimates, that is, $\hat{\beta} = (X'X)^{-1}X'y$, where X' is the transpose of X. The fitted model becomes $\hat{y} = X\hat{\beta} = X(X'X)^{-1}X'y = Hy$, where H is the hat matrix. The role of the hat matrix is to transform a vector of observed responses Y to the vector of fitted values $\hat{y}$. The least square estimates $\hat{\beta}$ are unbiased estimators of β

provided that the error terms ε_t are normally and independently distributed. The estimated error can be computed as $e = y - \hat{y} = (I - X(X'X)^{-1}X')y = (I - H)y$. If the Time Series data is large, H is computed using gradient descent.

$y = X\beta + \varepsilon$ where

$$
X = \begin{pmatrix} 1 & x_{1,1} & x_{1,2} & \cdots & x_{1,M} \\ 1 & x_{2,1} & x_{2,2} & \cdots & x_{2,M} \\ \vdots & \vdots & \vdots & \ddots & \vdots \\ 1 & x_{T,1} & x_{T,2} & \cdots & x_{T,M} \end{pmatrix}, y = \begin{pmatrix} y_1 \\ y_2 \\ \vdots \\ y_T \end{pmatrix}, \beta = \begin{pmatrix} \beta_0 \\ \beta_1 \\ \beta_2 \\ \vdots \\ \beta_M \end{pmatrix}, \varepsilon = \begin{pmatrix} \varepsilon_1 \\ \varepsilon_2 \\ \vdots \\ \varepsilon_T \end{pmatrix} \quad (6.7)
$$

Autoregressive model

The Multiple Linear Regression model can predict the variables of interest using a linear combination of predictors. The variables of interest can also be predicted using a linear combination of past values. This can be achieved using an autoregressive model (*AR*). Equation (6.8) presents an autoregressive model of order p, that is, $AR(p)$, where ε_t is white noise and c is a constant.

$$
y_t = c + \phi_1 y_{t-1} + \phi_2 y_{t-2} + \ldots + \phi_p y_{t-p} + \varepsilon_t = c + \sum_{i=1}^{p} \phi_i y_{t-i} + \varepsilon_t \quad (6.8)
$$

Moving average model

The moving average model (*MA*) is another model for predicting future values in a Time Series. Conceptually, the model is a linear regression that predicts the current value of the Time Series using the current and past errors (white noise) in a regression model. Equation (6.9) presents the moving average model of order q, i.e., $MA(q)$, where μ is the mean of the Time Series which is often assumed to be 0.

$$
y_t = \mu + \varepsilon_t + \theta_1 \varepsilon_{t-1} + \theta_2 \varepsilon_{t-2} + \ldots + \theta_q \varepsilon_{t-q} = \mu + \varepsilon_t + \sum_{i=1}^{q} \theta_i \varepsilon_{t-i} \quad (6.9)
$$

Autoregressive moving average model

The autoregressive moving average model (*ARMA*) for (weakly) stationary Time Series forecasting is obtained by combining the polynomials of the autoregressive model with the moving average model. Equation (6.10) presents the autoregressive moving average model with p autoregressive terms and q moving average terms, that is, $ARMA(p,q)$, where ε_t is white noise and c is a constant.

$$
y_t = c + \varepsilon_t + \sum_{i=1}^{p} \phi_i y_{t-i} + \sum_{i=1}^{q} \theta_i \varepsilon_{t-i} \quad (6.10)
$$

Autoregressive integrated moving average model

The autoregressive integrated moving average model (*ARIMA*) is a generalized ARMA forecasting model that can be applied to non-stationary Time Series. The model combines differencing with autoregression and a moving average model. A non-stationary Time Series can be transformed into a stationary one by using an initial differencing step once or multiple times. Thus, either the difference between two consecutive time points is computed, that is $y'_t = y_t - y_{t-1}$, or, if required, the second order differencing is computed, that is, $y^*_t = y'_t - y'_{t-1} = y_t - 2y_{t-1} + y_{t-2}$. Equation (6.11) presents the general form of the autoregressive integrated moving average model with p autoregressive terms, d integrated terms, and q moving average terms, that is, $ARIMA(p,d,q)$. Depending on the d term, y'_t may have been differenced more than one time. Furthermore, the predictors include both lagged values of y_t and lagged errors.

$$y'_t = \varepsilon_t + \sum_{i=1}^{p} \phi_i y'_{t-i} + \sum_{i=1}^{q} \theta_i \qquad (6.11)$$

The ARIMA model can be extended by adding a linear combination of seasonal past values and forecast errors, thus obtaining the SARIMA model (Seasonal ARIMA). Furthermore, if the Time Series exhibits long-range dependencies, the integrated parameter d can be fractional and the ARIMA model becomes an autoregressive fractionally integrated moving average model (Fractional ARIMA or FARIMA).

6.2.2.2 *Support vector machine*

Support Vector Machines (SVM) is an algorithm used for machine learning tasks such as classification and regression analysis. Support Vector Regression (SVR) is an extension of SVM used for estimating a function from observed data. In turn, the detected function is for conditioning the SVM. Given a Time Series $X = \{x_t | t \in \overline{1,T}\}$, the objective is to find a linear function $y = f(X) = X\beta + b$ or $y = f(X) = \phi(X)\beta + b$ in case the Time Series is not linear. To ensure that $f(X)$ is as flat as possible, we need to find the minimal norm value of $\beta'\beta$. We can formulate a convex optimization problem that minimizes eq. (6.12), where all the residuals are less then a value ε, i.e., $||y_t - (x_t\beta + b)|| \leq \varepsilon \ (\forall)t \in \overline{1,T}$.

$$J(\beta) = \frac{1}{2}||\beta||^2 \qquad (6.12)$$

For the non-linear case, we can use the kernel trick $k(X_i, X_j) = \langle \phi(X_i), \phi(X_j) \rangle$ that maps the kernel space X_i to a higher dimension future space $X_j = \varphi(X_i)$. After the kernel function is determined, we need to replace all dot products with the chosen kernel function and then proceed with the optimization problem as in the linear case.

Using eq. (6.12), there is a possibility that there is no function $f(X)$ that satisfies the constraints for all the points. We need to introduce two soft margins using the variables ξ_t and ξ_t^* to bound the regression errors. Equation (6.13) presents the objective function when the soft margins are employed, where C is a positive numeric value that controls the penalty imposed on the observations that lie outside the margin ε. The best solution is found when the loss function is minimal. Furthermore, using this formulation the following must be true:

- $y_t - (x_t\beta + b) \le \varepsilon + \xi_t \; (\forall)t \in \overline{1,T}$
- $(x_t\beta + b) - y_t \le \varepsilon + \xi_t^* \; (\forall)t \in \overline{1,T}$
- $\xi_t \ge 0 \; (\forall)t \in \overline{1,T}$
- $\xi_t^* \ge 0 \; (\forall)t \in \overline{1,T}$

$$J(\beta) = \frac{1}{2}||\beta||^2 + C\sum_{t=1} T(\xi_t + \xi_t^*) \tag{6.13}$$

Using Support Vector Machines (SVM) for predicting the Time Series has the following advantages [307] it,

- is not model dependent;
- is not dependent on linear, stationary processes;
- guarantees convergences to an optimal solution;
- has a small number of free parameters;
- can be computationally efficient.

6.2.2.3 Artificial neural network architectures

Artificial Neural Network architectures have also been used for Time Series prediction. Current research focuses on using Recurrent Neural Networks (RNN) architectures for the prediction task. Each RNN architecture is composed of different stacks of layers. Each layer contains multiple recurrent units. The most widely used recurrent units for Time Series prediction are:

- Recurrent Unit (RU) [110];
- Long Short-Term Memory (LSTM) [163] unit;
- Gated Recurrent Unit (GRU) [74].

These units can be used to create Neural Network architectures by stacking them together using different combinations. Depending on these combinations, we can design shallow architecture, that is, with a relatively small number of layers, or deep architectures, that is, with a large number of layers that contain processing units of different types [45].

Recurrent Unit

The Recurrent Unit (RU) computes the current hidden state using that of the previous time step as well as the current input. This is achieved through the use of feedback loops which connect the current RU state to the next state. Thus, these connections consider past information when updating the current cell state. Equation (6.14) presents the state updates of the RU unit, where:

- x_t is the Time Series data point;
- $z_t \in \mathbb{R}^d$ is the input and output of the cell at a time t;
- $h_t \in \mathbb{R}^d$ is the hidden state with a cell dimension of d;
- $W_i \in \mathbb{R}^{d \times d}$ and $V_i \in \mathbb{R}^{d \times d}$ are the weight matrices for the hidden state;
- $b_i \in \mathbb{R}^d$ is the bias vector for the hidden state;
- $W_o \in \mathbb{R}^{d \times d}$ is the weight matrix of the cell output;
- $b_o \in \mathbb{R}^d$ is the bias vector of the cell output;
- $\sigma \in [0, 1]$ is the the activation function of the gates, usually a sigmoid function, that is, $\sigma = \frac{1}{1+e^{-x}}$.

$$h_t = \sigma(W_i h_{t-1} + V_i x_t + b_i)$$
$$z_t = \tanh(W_o h_t + b_o)$$

(6.14)

Long Short-Term Memory

The Long Short-Term Memory (LSTM) unit has two components to its state: the hidden state and the internal cell state. The hidden state corresponds to the short-term memory component. The cell state corresponds to the long-term memory. LSTM avoids the vanishing and the exploding gradient issues. The cell also incorporates a gating mechanism which comprises of the input ($i_t \in \mathbb{R}^d$), forget ($f_t \in \mathbb{R}^d$), and the output gates ($o_g \in \mathbb{R}^d$). Equation (6.15) presents the state updates of the LSTM unit, where:

- x_t is the Time Series data point;
- $z_t \in \mathbb{R}^d$ is the input and output of the cell at a time t;
- $h_t \in \mathbb{R}^d$ is the hidden state with a cell dimension of d;
- $C_t \in \mathbb{R}^d$ is the cell state and $\tilde{C}_t \in \mathbb{R}^d$ is the candidate cell state at time step t which captures the important information to be persisted through to the future;

- $W_i, W_f, W_o, W_c \in \mathbb{R}^{d \times d}$ are the weight matrices of the input gate, output gate, forget gate, and the cell state;

- $V_i, V_f, V_o, V_c \in \mathbb{R}^{d \times d}$ are the weight matrices corresponding to the current input of the input gate, output gate, forget gate, and the cell state;

- $b_i, b_f, b_o, b_c \in \mathbb{R}^d$ are the bias vectors corresponding to the current input of the input gate, output gate, forget gate, and the cell state;

- $\sigma \in [0,1]$ is the activation function of the gates;

- $\odot$ is the element wise multiplication, i.e., Hadamard Product.

$$
\begin{aligned}
i_t &= \sigma(W_i h_{t-1} + V_i x_t + b_i) \\
f_t &= \sigma(W_f h_{t-1} + V_f x_t + b_f) \\
o_t &= \sigma(W_o h_{t-1} + V_o x_t + b_o) \\
\tilde{C}_t &= \tanh(W_c h_t + V_c x_t + b_c) \\
C_t &= i_t \odot \tilde{C}_t + f_t \odot C_{t-1} \\
h_t &= o_t \odot \tanh C_t \\
z_t &= h_t
\end{aligned}
\tag{6.15}
$$

Gated Recurrent Unit

The Gated Recurrent Unit (GRU) is a variant of a Recurrent Unit that simplifies the LSTM unit and improves performance considerably as it has only two gating mechanisms instead of three gates as in the case of LSTM. The GRU gates are i) the update gate ($u_t \in \mathbb{R}^d$), and ii) the reset gate ($r_t \in \mathbb{R}^d$). The update gate (u_t) is used as both the forget gate and the input gate. The reset gate (r_t) decides how much of the previous hidden state contributes to the candidate state of the current step [160]. The function $(1 - u_t)$ is used as an alternative as there is no forget gate. Furthermore, the GRU has only one state component, that is, the hidden state (h_t). Equation (6.16) presents the state updates of the GRU unit, where:

- x_t is the Time Series data point;

- $z_t \in \mathbb{R}^d$ is the input and output of the cell at a time t;

- $\tilde{h}_t \in \mathbb{R}^d$ is the candidate hidden state with a cell of dimension d;

- $h_t \in \mathbb{R}^d$ is the current hidden state with a cell of dimension d;

- $W_u, W_r, W_h \in \mathbb{R}^{d \times d}$ are the weight matrices of the update gate, reset gate, and the hidden state;

- $V_u, V_r, V_h \in \mathbb{R}^{d \times d}$ are the weight matrices corresponding to the current input of the update gate, reset gate, and the hidden state;

- $b_u, b_r, b_h \in \mathbb{R}^d$ are the bias vectors corresponding to the current input of the update gate, reset gate, and the hidden state;

- $\odot$ is the Hadamard Product.

$$
\begin{aligned}
u_t &= \sigma(W_u h_{t-1} + V_u x_t + b_u) \\
r_t &= \sigma(W_r h_{t-1} + V_r x_t + b_r) \\
\tilde{h}_t &= \tanh(W_h h_t + V_h x_t + b_h) \\
h_t &= u_t \odot \tilde{h}_t + (1 - u_t) \odot h_{t-1} \\
z_t &= h_t
\end{aligned}
\tag{6.16}
$$

6.2.3 Time series outlier detection

An outlier or an anomaly is a data point that significantly differs from other observations in a Time Series. Outliers can appear due to an experimental error or an anomaly in the measurement. Such suspicious points in the Time Series data must be identified and interpreted separately in order not to interfere with the analysis step and lead to wrong conclusions.

Anomalies have been classified into three groups [82], as follows: (i) Point Anomalies, (ii) Contextual Anomalies, and (iii) Collective or Pattern Anomalies.

Point Anomalies are a form of statistical noise, either produced by a wrong measurement or an event that can be of interest. Their main characteristic is that the Time Series returns to its previous state after just a few observations.

Contextual anomalies are data points or sequences that deviate from the Time Series expected pattern. These anomalies may fall within a range of expected values for the Time Series if analyzed independently from the other points. In practice, these anomalies are considered a deviation from the norm.

Collective or Pattern anomalies appear in groups. The data points that belong to the anomaly taken individually may or may not be outliers. These anomalies are a collective of observations which are outliers with respect to the rest of the Time Series data.

There are multiple machine learning techniques used for outliers detection [146, 82] regardless of the anomaly type. These methods are grouped into the following categories:

- Supervised methods, e.g., Regression models, one class SVM, Recurrent Neural Networks, k-nearest neighbour, Isolation Forests, and more;

- Unsupervised methods that include clustering algorithms, for example, DBSCAN, k-Means, k-Medoids, Expectation-Maximization, and others, and neural networks, for example, Self-Organizing Maps.

Supervised methods are used for predictions using historical data. If the difference between the predicted value and the real value is too high, then that value can be interpreted as an anomaly. The methods presented in Subsection 6.2.2 can be used for outlier detection.

Unsupervised methods use clustering algorithms to group similar data points. The points that fall outside the clusters with a majority of points or those which create clusters further away from the majority clusters are considered outliers. The methods employ different approaches to detect clusters with anomalies, starting with simple metrics based on intra-cluster and inter-cluster distances (for example, k-Means, k-Medoids), measures that aggregate the data using hierarchies (for example, BIRCH, CURE), density (for example, DB-SCAN, CLARANS, etc.) or distribution (for example, Gaussian Mixture Models, Expectation-Maximization).

6.2.4 Time series change point detection

The problem of change point detection in Time Series data deals with finding the point in time when the properties (mean, variance) of the Time Series change abruptly [18]. The Change Point Detection technique contains three components [357]: (i) the constraints on the number of change points, (ii) the cost function, and (iii) the search method.

A change point is a transition point between different states in the Time Series data. A continuous state is defined as a segment. For a given Time Series X and a set of indexes $\tau = \{t_1, t_2, ..., t_K\} \subseteq \{1, 2, ..., T\}$, a segmentation is defined by $\{X\}_{t_k}^{t_{k+1}}$ or simply $X_{t_k, t_{k+1}}$, where the dummy indexes t_0 and t_{K+1} are implicitly available and K is the number of change points. Using this notation, the entire Time Series is $\{X\}_{t_1}^{t_T}$. The change point detection methods are dependent on prior knowledge about the number of change points K, thus they are grouped into two categories: (i) supervised (K is known), and (ii) unsupervised (K is unknown).

The change point detection problem is formulated as detecting the best possible segmentation τ for a Time Series X according to a criterion (total cost) function $V(\tau, X)$. The criterion function is the sum of costs $c(\cdot)$ of all the segments $X_{t_k, t_{k+1}}$ that define the segmentation $V(\tau, X) = \sum_{k=0}^{K} c(X_{t_k, t_{k+1}})$. The "best segmentation" $\hat{\tau}$ is the minimum of the criterion function $V(\tau, X_t)$. Depending on the two categories, that is, supervised and unsupervised, the entire problem is reduced to solving an optimization problem. For a known number of change points K, the change point detection problem consists in solving $\min_{|\tau|=K}(V(\tau, X_t))$, while for an unknown number of change points it consists in solving $\min_{\tau}(V(\tau, X_t) + \delta(\tau))$ where $\delta(\tau)$ is a measurement that balances out $V(\tau, X_t)$ over-fitting when small change amplitudes are detected. Therefore, the search method is how the discrete optimization problem is solved and the constraint is defined by the number of change points to detect.

In the literature, the cost functions were grouped in two categories: (i) parametric models, and (ii) non-parametric models.

The parametric models include:

i) Maximum likelihood estimation, where $V(\tau, X_t)$ is equal to the negative log-likelihood [147];

ii) Multiple linear model, where $V(\tau, X_t)$ is equal to the sum of squared residuals [289, 151];

iii) Mahalanobis-type metric, where $V(\tau, X_t)$ is equal to the sum of cost functions extended through the use of Mahalanobis-type seminorm [90].

The non-parametric models include:

i) Non-parametric maximum likelihood, where the cost function is computed using an empirical cumulative distribution function [400].

ii) Rank-based detection [233] uses statistical inference to replace the data samples by their ranks within the set of pooled observations;

iii) Kernel-based detection maps the original Time Series data onto a reproducing associated space that is associated with a user-defined kernel function [155].

The search methods are also grouped into two categories. The first category, that is, optimal detection, contains models for finding the exact solution to the optimization problem. The second category, that is, approximate detection, solves the optimization problem yielding an approximate result. The approximate detection is used when it is desired to reduce the computational complexity.

When the number of change points is known then the optimal solution can be computed using the exact segmentation dynamic programming model (OPT) [296]. The OPT algorithm solves the problem recursively by using the additive nature of the objective function $V(\tau, X_t)$. Under these observations the optimal partitions with $K - 1$ elements of all sub-series of X_t are known, thus the first change point of the optimal segmentation can be computed. Used recursively, the complete segmentation is then computed.

When the number of change points is unknown then the optimal solution can be computed using exact segmentation to minimize the penalization function. The model Pruned Exact Linear Time (PELT) [200] finds the exact solution when the penalty is linear. PELT considers each Time Series data sample sequential. The explicit pruning rule tests if each sample point and determines if it is a potential change point. PELT works under the assumption that each region's length is randomly drawn from a uniform distribution.

Approximate Detection methods are used when the computational complexity of optimal methods do not provide timely results. The literature provides three major classes of approximation detection methods: (i) Window-based change point detection [225], (ii) Binary segmentation [321, 127], and (iii) Bottom-up segmentation [199].

The window-sliding algorithm [225] computes the discrepancies between two adjacent windows that slide along the data point of a Time Series X_t. A discrepancy is a value calculated for each index of time t between the immediate past, in the left window, and the immediate future, in the right window. Using this method, the algorithm detects peaks, that is, large values, when two windows cover dissimilar segments and compute the discrepancy curve. The change point indexes are determined using a peak detection procedure on the curve.

Binary segmentation [321, 127] is a greedy sequential algorithm that searches for the change point that minimizes the sum of costs. The Time Series is then split into two at the index of the determined first change point and restarts the computation on the new sub-series. The algorithm stops when either the required number of change points are found or when a stopping criterion is met.

Bottom-up segmentation [199] is a sequential approach used to perform fast signal segmentation that starts with many change points and successively deletes the less significant ones. Thus, the time series is divided in many segments along a regular grid and then, the contiguous segments are successively merged according to a measure of how similar they are. Moreover, bottom-up segmentation can extend any single change point detection method to detect multiple changes points.

6.3   NoSQL databases

The term NoSQL was first used to describe relational databases that did not use the SQL language to interact with the data. The term was picked up again by the advocates of non-relational databases that promoted databases that did not adhere to the relational algebra for modeling data. The main reason for the NoSQL movement was to seek alternatives for solving problems for which relational databases are a bad fit.

Some of the motives for using new models for data management and developing new NoSQL databases are [337]:

1. *Avoidance of unneeded complexity*. Relational databases provide strict data consistency and ACID (Atomicity, Consistency, Isolation, Durability)

transactions which are restrictive for some applications. These relational database features might be more than necessary for particular applications and used cases.

2. *High throughput*. The strict consistency and ACID properties restrict the throughput for relational databases. NoSQL uses looser consistency, and less restrictive transaction properties, and different storage methods and data persistence models than relational databases to improve throughput. Moreover, NoSQL databases' processing capabilities are improved using distributed and parallel algorithms, that is, MapReduce.

3. *Scaling data*. Due to the huge volume of data, new technologies are needed to store and process this data. NoSQL databases try to address the following current problems that relational databases have: (i) scaling out data while respecting the ACID properties for transactions, (ii) low performance of data processing as the volume of information increases, and (iii) rigid schema design that makes it difficult to share data properly. NoSQL databases are designed to: (i) easily scale out as new nodes can be added and removed without causing the same operational efforts to perform shading as in relational database cluster-solutions, (ii) not rely on highly available hardware, (iii) use flexible schema-less design to adapt to the needs of sharding and scaling data, and (iv) lower the costs of scaling.

4. *Avoidance of Expensive Object-Relational Mapping*. NoSQL databases are designed to store data using formats that can easily map onto the data structures available in high-level programming languages, thus avoiding expensive object-relational mapping and lowering the complexity of interacting with the data stored in the database.

5. *Requirements of Cloud Computing*. NoSQL databases meet the major requirements needed by cloud computing environments [143]: (i) high ultimate scalability (especially in the horizontal direction), (ii) elasticity to handle the application fluctuations in their access patterns, (iii) ability to run on commodity heterogeneous servers, (iv) fault-tolerant to multiple unforeseen errors, and (v) low administration overheads.

6.3.1 Transaction properties

ACID (Atomicity, Consistency, Isolation, Durability) is a set of database transaction properties that ensure data validity in the case of system failure. Atomicity requires that each transaction is either committed or rolled back, thus, if one part of the transaction fails, then the entire transaction fails, and the database state is left unchanged. Consistency ensures that each transaction, successful or not, will preserve a valid state of the database. Isolation ensures that the final state of the system after the execution of concurrent transactions could be obtained if the transactions were executed serially. Durability ensures that the modifications executed by a committed transaction remain persistent even in case of a system failure.

BASE (Basically Available, Soft-state, Eventual consistency) is a set of database transaction properties which favor availability over consistency operations. The Basically Available constraint states that the system does guarantee the availability of the data as defined by the CAP Theorem. A request will always have a response. A 'failure' response can be obtained if the requested data is in an inconsistent or a changing state. A Soft-state system could change over time due to committed transactions that were not made persistent. Eventual consistency guarantees that the system will end up in a consistent state even if the consistent state of each transaction is propagated on all the nodes.

6.3.2 CAP theorem

A distributed database system can only provide two of the three following characteristics: Consistency, Availability, and Partition Tolerance. These characteristics are known as the CAP Theorem. In the Big Data world, the CAP Theorem is an essential marker for determining the required characteristics of applications. Furthermore, when they need to make trade-offs between the three, to ensure the functionality and high performance of unique use cases is required. Consistency describes if and how a system remains in a consistent state after the execution of an operation. A distributed system is typically considered to be consistent if after an update of some writer's operation all readers see its updates in some shared data source. A high availability especially ensures that a system is designed and implemented in a way that allows it to continue operation (that is, allowing reading and writing operations) even if nodes in a cluster crash or some hardware or software parts are down due to upgrades. Partition Tolerance is understood as the ability of the system to continue operation in the presence of network partitions (temporary or permanent) which cannot connect to each other. Sometimes, partition tolerance is described as the ability of a system to continue functioning while nodes are added or removed dynamically. The main traits of the characteristics of the CAP Theorem (Table 6.1) are:

1. Use 2-phase commit, a protocol that ensures the Atomicity ACID property for transactions in a distributed system.

2. Use cache validation protocols, which ensure the uniformity of shared resource data stored on multiple nodes in their local caches.

3. Make minority partitions unavailable. A minority partition is a partition composed of nodes fewer or equal to half of the total number of nodes.

4. Use pessimistic locking, which locks a resource from the time it is first accessed by a transaction until it has either been committed or rolled back.

5. Use optimistic locking, which makes a resource available to other concurrent transactions by not locking it when it is first accessed by a transaction.

6. Conflict resolution, which permits concurrent transactions to change the same data at multiple sites in a distributed environment.

Table 6.1: CAP theorem traits.

Characteristics	Traits	Databases
Consistence + Availability	2-phase commit (2PC) Cache-validation protocols	Oracle, PostgreSQL, MySQL, MS SQL Server
Consistence + Partition Tolerance	Unavailable minority partitions Pessimistic locking	MongoDB, HBase, Redis
Availability + Partition Tolerance	Optimistic locking Conflict resolution	CouchDB, Cassandra, Riak

6.3.3 Distributed database architectures

A distributed architecture is a model in which resources located on network computers are put together in a resource pool to achieve a common goal. One of the most important features of a distributed system is its ability to scale, and to handle a growing amount of work. Scalability can be:

1. Vertical (scale up/down) meaning that the system can be scaled by adding more resources to an existing machine, for example, CPU, RAM, HDD.

2. Horizontal (scale out/in) meaning that the system can be scaled by adding more machines into the resource pool, that is, adding nodes to the cluster.

Horizontal scaling is either homogeneous, that is, all the nodes are the same, or heterogeneous, that is, nodes have different configurations.

A cluster is a set of tightly connected computers that work together and share resources among them using one of the following:

1. A shared-everything architecture is a distributed computing architecture in which each node adds its resources to the resource pool. More specifically, the user sees the total amount of memory, CPU or disk storage and not individual capacities.

2. A shared-nothing architecture is a distributed computing architecture in which each node is independent and self-sufficient. More specifically, none of the nodes share memory, CPU or disk storage.

6.3.3.1 Replications methods

A distributed system must ensure consistency between redundant resources and needs to improve reliability, fault-tolerance, and accessibility. These conditions are achieved through replication, which enables the storage of the same data on multiple nodes. Replication can be seen from an architecture perspective and a transaction perspective.

From an architecture perspective, there are two types of replications, Primary-Secondary and Multi-Primary. When using a Primary-Secondary replication architecture, a transaction can be started on the primary nodes, which in turn support all create, read, update, and delete (CRUD) operations from clients. When using this replication type, the secondary nodes support only read operations from clients. When using a Multi-Primary replication architecture, a transaction can be started on any of the nodes, and each node supports all the CRUD operations.

From a transaction perspective, there is synchronous and asynchronous replication. Synchronous replication guarantees 0 data loss, that is, a transaction is either completed on all the nodes or not at all. A transaction is not considered complete without acknowledgment from all the nodes. When using synchronous replication, the overall performance decreases considerably because applications wait for write transactions to complete before proceeding with further work. Asynchronous replication considers that a transaction is complete as soon as it is committed locally, and the transaction is triggered on the remote nodes where the modifications are committed with a small lag. When using asynchronous replication, the overall application performance is increases significantly but, in case of failure, it does not guarantee an updated copy of the data on all the nodes.

6.3.3.2 Data partitioning methods

Data partitioning in a distributed database is done using fragmentation. Fragmentation implies splitting a relation R in n fragments R_i, $i = \overline{i,n}$. Partitioning data includes different modeling techniques (Fig. 6.1): horizontal, vertical, mixed vertical-horizontal, and mixed horizontal-vertical.

When using horizontal fragmentation (Fig. 6.1(a)), the original relation R can be reconstructed from the fragments R_i by using the UNION operation, thus $R = \bigcup_{i=1}^{n} R_i$. The JOIN operation is used to reconstruct the original relation R when using vertical fragmentation (Fig. 6.1(c)): $R = R_1 \bowtie R_2 \bowtie \ldots \bowtie R_n$.

When using mixed vertical-horizontal fragmentation (Fig. 6.1(b)), the original relation R is first split into R_i fragments using vertical fragmentation $(R = R_1 \bowtie R_2 \bowtie \ldots \bowtie R_n, i = \overline{i,n})$, and then each fragment R_i is split using horizontal fragmentation into R_{ij} $(R_i = \bigcup_{j=1}^{m_i} R_{ij})$. To reconstruct the relationship, R first we need to apply the UNION operation on the fragments and then the

JOIN operation on the result of the union, that is, $R = \bigcup_{j=1}^{m_1} R_{1j} \bowtie \bigcup_{j=2}^{m_2} R_{2j} \bowtie$
$\dots \bowtie \bigcup_{j=1}^{m_n} R_{nj}$.

Finally, when using the mixed vertical-horizontal fragmentation (Fig. 6.1(d)), the original relation R is first split into R_i fragments using vertical fragmentation ($R = \bigcup_{i=1}^{n} R_i, i = \overline{i,n}$), and then into R_{ij} ($j = \overline{i,m_i}$) fragments using horizontal fragmentation ($R_i = R_{i1} \bowtie R_{i2} \bowtie \dots \bowtie R_{im_i}$). To reconstruct the original relation R, first we need to apply the JOIN operation on the fragments and then the UNION operation on the result, i.e., $R = \bigcup_{i=1}^{n} (R_{i1} \bowtie R_{i2} \bowtie \dots \bowtie R_{im_i})$.

Sharding is a horizontal partitioning of the data in a database widely used by NoSQL databases.

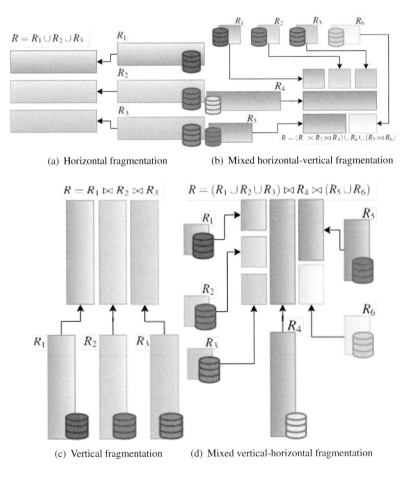

(a) Horizontal fragmentation

(b) Mixed horizontal-vertical fragmentation

(c) Vertical fragmentation

(d) Mixed vertical-horizontal fragmentation

Figure 6.1: Data fragmentation methods

6.3.4 Types of data

The data types stored by the different database technologies are often described by the data model used and are: structured, unstructured and hybrid.

Structured data refers to data that has a predefined data model and has a strict schema that generally adheres to the Relational Algebra Theory. The dataset's attributes are predefined and have a strict data type: number, datetime, string, to name a few. Structured data technologies are Relational Databases, Data Warehouses, and Data Marts, Enterprise Resource Planning software (ERP), and Customer Relationship Management software (CRM).

Unstructured data refers to data that does not have a predefined data model and does not adhere to a strict schema, and generally, it uses a flexible schema-free model. The dataset's attributes are not predefined and they do not have a strict data type. Unstructured datasets also employ the use of different data structures, for example, sets, lists, maps. Technologies for unstructured data are NoSQL databases, distributed storage, and processing frameworks, data Lakes (a storage repository that stores vast amounts of raw data in its native format), and Enterprise Content Management Systems (CMS) that manage the complete life cycle of content.

Hybrid Big Data combines structured and unstructured data types, data sources and correlates the information, and the storage technologies.

The sources for gathering the data, regardless of type, are either machine-generated (data that is created by a machine without human intervention) or human-generated (data that is generated by the interaction of humans with computers).

6.4 NoSQL time series database

A Time Series Database (TSDB) is a database optimized to work with Time Series data. Currently, there are multiple Database Management Systems that offer Time Series data storage, processing, querying, and analysis functionalities. In this sections, we discuss some of the most well-known and well-established databases of this kind.

6.4.1 InfluxDB

InfluxDB [172] is an open-source, schema-less Time Series database written in programming language GO that provides a SQL-like query language called InfluxQL (Influx Query Language). InfluxDB can run on the Cloud or the premise using a single machine or on pseudo-distributed and distributed environments.

6.4.1.1 Architecture

InfluxDB uses a Time-Structured Merge Tree (TSM) for persistently storing in a columnar format sorted and compressed Time Series data. The TSM is similar to a Log-structured merge-tree (LSM). To provide atomicity and durability, the storage engine utilizes write-ahead logging (WAL) and a collection of read-only data files with a physical structure similar to SSTables. Each read-only data file is handled by the FileStore which mediates access to all TSM files on disk. The FileStore has the following sections: header, blocks, index, and footer.

InfluxDB supports data sharding. Depending on the retention policy, a shard is created for each block of time. Each shard maps to a separate storage engine database containing its own WAL and TSM files.

To improve performance, InfluxDB uses the Cache to store all the WAL data in-memory. For a write operation, each new record is appended to the WAL file and added to the Cache. Update operations appear as normal writes because cached values are overwritten, while deletes are handled by writing a new entry in the WAL file followed by updating the data in the Cache and FileStore.

InfluxDB handles queries as follows: (i) searches the data files to match the time interval to the query, (ii) sequentially reads the TSM index to retrieve the location of the block, and (iii) decompresses the block and extracts the specific record. To improve performance, InfluxDB permits Time Series index (TSI) and columnar indexes by tags using inverted indexes.

Concurrency control is achieved using subscriptions. Data is written to InfluxDB through subscriber endpoints [172]. The number of writers is set using the system parameter *write-concurrency*.

Influx Query Language (InfluxQL) is an SQL-like query language used for manipulating and querying the records stored within the InfluxDB database.

6.4.1.2 Data model

When working with InfluxDB, a **time** column is defined for every record in the database. The rest of the attributes of a record are named **fields**. As in similar NoSQL Key-Values or Document-Oriented Databases, each **field** is defined by a **field key**, that is, the name of the attribute, and the **field value**, that is, the actual data that is always associated with a time stamp. The **field value** has a data type, for example, string, float, integer. All the pairs **field keys** and **field values** make up a **field set**. Fields are not indexed, thus queries that filter the data using **field values** must pass over all the records in the database. Besides **fields**, InfluxDB also has tags which represent metadata about the records stored in the database. As **fields**, **tags** are composed of **tag keys** and **tag values**. A **tag set** is a combination of all the **tag key** and **tag value** pairs. Although tags are optional, they play an important role when querying the data because they are indexed improving the performance of read statements that filter the data using metadata information.

The container for **time**, **labels**, and **tags** columns is called a **measurement**. The **measurement** is conceptually similar to a table in a relational database. As in relational databases, the basic functionality of the **measurement** is to better describe the stored data. A **measurement** is defined in an InfluxDB **database** and can belong to multiple retention policies. A database is a logical container for all the objects defined to store and manipulate the Time Series data. A **retention policy** describes the number of replicas stored in the cluster and when the data expires. By default, the **retention policy** is **autogen** with an infinite duration and a replication factor of 1.

A **series** is defined in InfluxDB as a collection that shares a **retention policy**, a **measurement**, and a **tag set**. A single data record in the Time Series is called a **point**. It is defined by its series and timestamp, and has four components: a measurement, tag set, field set, and a timestamp.

InfluxDB's data model does not support the JOIN operation.

6.4.2 Kdb+

Kdb+ is a high-performance multi-core process and multi-threading Time Series database developed to store and manipulate efficiently high-volumes of real-time and historical time data. Kdb+ uses a column-based persistence model with in-memory capabilities [194]. Kdb+ uses the **q** language to query, process, and enable analysis directly at the database level [207].

6.4.2.1 Architecture

Kdb+ uses a column-oriented data persistence model. Records in this model are stored using column vectors instead of row vectors. Thus, a table is first split using vertical fragmentation, each resulting relation storing the values for each column. At the physical level, Kdb+ can use *splayed tables* to store each column of a table in a separate file. A keyed table, that is, a dictionary mapping a table of key records to a table of value records, cannot be splayed. When dealing with Time Series data, this approach is useful for grouping operations and data aggregation. As the information is stored by column, the grouping operation will read from the persistence storage device sequences of information, improving retrieval performance. Aggregation functions are applied to column values, either grouped or otherwise. The column-oriented persistence model improves the computation time of such functions as the data is read from one location while the performance time scales with the full volume of data. Kdb+ supports replication mechanisms for data reliability.

Using the column-oriented data persistence model, Kdb+ ensures the following three features [5]:

i) data predictability: the Kdb+ database management system can predict exactly where the data can be found as it is of fixed length;

ii) data locality: the Kdb+ database management system has a high probability to hit or block the data cache;

iii) improved read operations: the Kdb+ database management system improves the read performance as it only retrieves the data required by the queries.

The Kdb+ management system uses an in-memory data storage [116] and manipulation features aimed at improving query performance and data retrieval. This architectural design was employed to bring the data closer to the processing model. To maniputete the data, Kdb+ uses the **q** programming language. The **q** language performs operations in-memory and, to improve data manipulation, it is designed to also store the tables using an in-memory data storage model.

Kdb+ uses deterministic concurrency control through the use of partition-based timestamp ordering. When a transaction starts it is marked with a timestamp. The execution is done in the order of the transaction timestamp on each partition [207].

The **q** programming language is a vector processing language [56], thus all the records stored in Kdb+ are ordered using vectors of ordered lists. Queries written in **q** do not require the use of slow cursors as in the case of other SQL base languages when dealing with the Time Series data. It provides unique Time Series JOIN operations (for example, WINDOW JOIN), for improving the data processing and retrieval time. To further improve performance, Kdb+ supports primary and secondary indexing.

6.4.2.2   Data model

Kdb+ uses the relational data model. Information is modeled using tables or keyed tables. A keyed table is a dictionary where the key is a table of unique key records and the value is a table of value records [56]. This representation improves lookup performance, as search using dictionaries has a time complexity of $O(1)$.

The relationship between tables is done using foreign keys and link columns. A foreign key is similar to the ones in traditional relational databases. A link column is similar to a foreign key but the lookup must be performed manually [56].

Kdb+ data model defines all the relational algebra operators. Furthermore, it provides special JOIN operations to deal with Time Series data, such as WINDOW

JOINs. A WINDOW JOIN [201] matches two Time Series that share a common key and lie in the same time window.

6.4.3 *Prometheus*

Prometheus is a Time Series database developed by SoundCloud [57]. This database development was driven by the necessity for building a new metric-based monitoring system that lets users analyze their applications and infrastructure performance in real-time [57]. PromQL (Prometheus Query Language) is the query language used for interacting, selecting, and aggregating the Time Series data stored in Prometheus in real-time.

6.4.3.1 Architecture

Prometheus architecture is composed of multiple modules. The core modules are integrated into the Prometheus Server that contains three components: (i) the Retrieval component, (ii) the Time Series Database, and (iii) the HTTP Server.

The Retrieval component contains two modules, that is, Service Discovery and Scraping.

The Service Discovery module helps Prometheus integrate many service discovery mechanisms, such as, Kubernetes, EC2. This module provides the tools that Prometheus needs to determine if the environment is working properly. The Service Discovery module provides the list of monitored applications.

The Scraping module receives the list of applications that need to be monitored by the Service Discovery module. To gather metrics from the applications, Prometheus uses HTTP data gathering requests called **scrape** [57]. The response for a request is parsed and stored into the Prometheus storage. The Scraping module uses the Client Libraries and the Exporters to request data and retrieve metrics from either in-house developed or third party applications.

The Client Libraries module provides the connection tools, monitoring packages, and metrics features for developing in-house applications in many programming languages. Through these libraries, the Scraping module can gather application metrics and store them in the database. The concurrency control is achieved through the use of **query.max-concurrency** parameter that sets the maximum number of queries executed concurrently.

The Exporter is a software utility used to extract metrics for applications. The Exporter receives data gathering requests from the Scraping module, transforms them into the application known format, and returns the required measures to the Prometheus database.

The metrics gathered through the Scraping module are stored in the Prometheus Time Series Database. The database is not distributed, everything is stored locally. To improve data management, the records stored in the database are compressed. The Long-Term Storage (LTS) model is similar to the one proposed for Gorilla Time Series database [282]. Data durability is ensured by write ahead logging (WAL). The storage engine utilizes LevelDB [97] for indexing data.

The HTTP Server component is used to expose Prometheus metrics to the external components. Prometheus uses the Alertmanager to push alerts as notifications using different means, such as, e-mail, chat applications.

Prometheus also provides Dashboards to permit users to interact directly with both raw data and evaluates PromQL queries. To improve the search and data retrieval, inverted indexes are supported.

6.4.3.2 Data model

Prometheus uses a multi-dimensional data model, where each Time Series is defined by a **metric name** and a set of key-value dimensions [285]. Temporary Time Series can be generated by querying the data using PromQL.

Prometheus data model uniquely identifies a Time Series through the use of a **metric name** and optional **labels**. The **metric name** is a descriptive attribute used to characterize the measured feature of a target application.

The **labels** are the base of the multi-dimensional model used in Prometheus. Each **label** represents a characteristic of the **metric name**. A dimensional instance for a **metric name** is defined as a combination of **labels**. PromQL uses the dimensions to select, filter, and aggregate the Time Series data.

6.4.4 OpenTSDB

OpenTSDB is a distributed and scalable Time Series database built on top of HBase. It achieves replication and horizontal scale through HBase's use of the Hadoop File System (HDFS). OpenTSDB provides its own query language to access, manage, and manipulate the Time Series data.

6.4.4.1 Architecture

OpenTSDB architecture consists of three components: (i) HBase for efficiently managing data, (ii) the Time Series Daemon (TSD), and (iii) multiple command line utilities. HBase is a distributed, salable wide-column database [347].

OpenTSDB uses multiple TSDs, each running independently, to collect data from different sources. TSDs do not share the state among them and each uses HBase for data storage and retrieval. Data collectors located at the target side send data using the TSD RPC protocol [100]. The daemon is responsible for matching where to append the new data in the Time Series database.

OpenTSDB uses Immediate Consistency (IC) as its consistency model. IC requires that a transaction is an all or nothing operation. Thus, if anything goes wrong, the entire transaction is rolled back.

OpenTSDB concurrency control mechanism permits concurrent write operations without using a table or row locks. In case a transaction is rolled back, OpenTSDB avoids multiple writers by making records idempotent. Idempotency is achieved by enforcing a fixed timestamp that always forces the writer process to write data at the appropriate row instead of creating a duplicate record. Furthermore, records are stored using a timestamp ordering.

OpenTSDB query language provides CRUD and aggregation commands. Indexing is provided through the HBase indexing functionalities.

6.4.4.2 Data model

OpenTSDB stores data using a Time Series model. In OpenTSDB, as in the case of Prometheus, a Time Series contains a time ordered sequence of **values** for a particular **metric**. The OpenTSDB data model defines a Time Series as a collection of data points, where each data point is a pair that contains the **metric**, the **timestamp**, the **value** for the **metric** at a given **timestamp**, and one or more **tags**. The **tags** are descriptors that characterize the **metric**. The TSD stores the **timestamp** for each **metric** in HBase, the **value**, and the **tags**.

Filtering the data is done using **tags**, while aggregation is done on the **value**. Grouping can be achieved using the **tags**. OpenTSDB provides **downsampling** operators to reduce the number of data points returned by intervals and methods. The querying engine does not provide JOIN operations.

6.4.5 TimescaleDB

TimescaleDB [354] is a scalable Time Series database optimized for large workloads and fast query performance. TimescaleDB incorporates both the relation and NoSQL models. TimescaleDB data is stored in tables and provides full SQL support. To address some limitations of the relational model, TimescaleDB optimized data flexibility by proving both wide-table and narrow-table design to address specific use-cases.

6.4.5.1 Architecture

TimescaleDB is built to run within PostgreSQL as an extension to this relational database. Using this architectural design, TimescaleDB takes advantage of all the features available out of the box in PostgreSQL such as reliability, security, client connectors, replication. TimescaleDB adds its own layer of functionalities and models on the top of PosgreSQL when it comes to query planner, data model, and execution engine. TimescaleDB also provides horizontal scaling.

The data model employed by TimescaleDB is a singular table called a **hyper-table**. The **hypertable** is a table that stores continuous time and space intervals. Users can use the SQL language to define the data model as well as to manipulate, retrieve, and visualize the data. A **hypertable** is defined at the schema level and stores data into columns with a data type. Each **hypertable** has at least one column used for storing the time values. As an option, another column can be used to store the values that represent the *partitioning key*. A TimescaleDB schema can contain multiple **hyper-tables**.

To improve query response time, the time and partitioning key columns are automatically indexed. All the PostgreSQL index types can be created for indexing other columns belonging to a **hypertable**.

Concurrency control and logging are achieved through the mechanism already in place in PostgreSQL, such as, Multi-version Concurrency Control (MVCC) and write ahead logging (WAL).

6.4.5.2 Data model

The narrow-table model implies that each metric attribute (tag) and each possible metric value, resulting in a metric/tag-set combination, is an independent Time Series. This model stacks information for the same data point within two columns: one containing the context (the attribute name) and another storing the value. Thus, the Time Series data scales with the cross-product of the values of each tag (the tag cardinality). This model is useful for use-cases when each metric is collected independently. The schema flexibility provided by TimescaleDB permits the addition of new tags as the data is collected. The downside of this model is that its performance decreases when collecting metrics with the same timestamp because the timestamp attribute will contain duplicates which in turn will increase the number of records and table size. Moreover, the number of JOIN operations will increase and the queries will become more complex when trying to retrieve correlated metrics.

The wide-table models stores unstacked data, that is, for each data point a column stores the values for different attributtes. The main advantage of the wide-table model is that queries that correlate multiple metrics do not require the JOIN operations as each metric is defined by the same timestamp which in turn is a record attribute. This format is similar to the one found in a relational database, which allows the preservation of relationships within the records.

Data stored in TimescaleDB can be normalized as foreign keys are supported, thus permitting the storage of metadata in secondary tables decreasing the effort of managing data. By allowing data normalization, TimescaleDB improves the INSERT operations, decreases the storage costs by removing duplicates, and minimizes update mappings. Furthermore, the JOIN operation can be used to correlate data stored in different tables at query time.

6.4.6 Comparison

Table 6.2 presents a summary comparison of the presented Time Series databases. In this comparison, we will look explicitly at the following features:

i)   The ability of the database to use horizontal scaling by adding new nodes. This is achieved only by OpenTSDB and TimescaleDB through the use of HBase and PostgreSQL, respectively.

ii) The ability to replicate data. Except for Prometheus which stores data locally without replicating the data files, all the other discussed Time Series databases can replicate data locally or on a cluster.

iii) Data reliability is ensured by all the discussed databases using write ahead logging (WAL), except OpenTSDB which uses Immediate Consistency.

iv) The ability of multiple writer processes to write data concurrently is achieved using different techniques.

v)   The query language is also an important factor to be taken into account when choosing a database. Except for TimescaleDB which uses SQL, all the databases provide their own query language.

vi) The similarity of a query language to SQL might influence the adoption of a database management system. Only OpenTSDB uses a query language different in syntax from SQL.

vii) All the databases provide CRUD operations.

viii) Filtering is achieved by all the databases.

ix) Aggregation and grouping are available in all the databases.

x)   JOIN operations are present only in Kdb+ and TimescaleDB. Kdb+ also implements Time Series specific JOINs.

xi) Primary Indexing is supported by all the databases. The primary indexing is usually done on the timestamp.

xii) Secondary Indexing is supported by all the databases, except for Prometheus. Secondary indexes can be added on the time field or the other fields stored in the database.

xiii) Persistent storage is provided by all the discussed databases.

Table 6.2: Time series database comparison.

Criterion	InfluxDB	Kdb+	Prometheus	OpenTSDB	TimescaleDB
Horizontal scaling	no	no	no	yes	yes
Replication	yes	yes	no	yes	yes
Reliability	WAL	WAL	WAL	IC	WAL
Concurrency control	subscriptions	deterministic	parameterized	timestamp	MVCC
Query Language	InfluxQL	q	PromQL	own language	SQL
SQL-like	yes	yes	no	no	yes
CRUD	yes	yes	yes	yes	yes
Filtering	yes	yes	yes	yes	yes
Aggregation	yes	yes	yes	yes	yes
JOINs	no	yes	no	no	yes
Primary Indexing	yes	yes	yes	yes	yes
Secondary Indexing	yes	yes	yes	no	yes
Storage	own	own	own	HBase & HDFS	PostgreSQL FS

6.5 Big data environment for time series

Recent advances in networking, caching, and computing have significant impacts on the development of smart cities, agriculture, healthcare, to mention a few. The development of smart applications is heavily influenced by information and communications technology and broth innovations in various areas, including data analysis, management, and storing for Time Series and data collected from smart devices in the Internet of Things (IoT) [158]. IoT enables the collection of Time Series data thought the use of smart devices that incorporate sensors.

Big Data analytics applications require optimization, localization, and globalization enabled through the use of Edge Computing, Fog Computing, and Cloud Computing, respectively. Edge computing is an architecture that enables the optimization of data processing and analysis at the edge of the network on embedded systems and smart devices. Fog Computing is an architecture that permits the localized decentralization of the data gathering, analysis, and storage processes on dedicated physical or virtual devices that use direct communication between them. Cloud Computing facilitates globalization by offering on-demand availability of computer system resources. These three distributed computer architectures provide the tools for:

- collecting and analyzing data in real-time on the embedded systems and sensors from the Edge Computing nodes;

- providing data storage, management, visualization, and more advanced analysis of data on the Fog Computing environments;

- facilitating Big Data Analysis and Data Warehousing for business intelligence processes and historical data storage using Cloud Computing.

Figure 6.2 presents the entire environment for collecting, storing, and analysing Time Series. Machine learning algorithms should be implemented at different layers depending on the type of analytics and where it is done.

Applications at the Edge Computing layer can incorporate pre-trained models for forecasting, anomaly detection, and change point detection. At this layer, data can be stored using data management systems that require modest computing capabilities, such as Prometheus. The storage systems and data analysis models, enable applications to alert users about system and functionality changes by reacting in real-time.

At the Edge computing layer, the data is collected from IoT smart devices and sensors. This data is in the form of Time Series and can contain information from different providers:

- consumer applications, such as, smart homes, elder care systems;

- commercial applications, such as, medical and healthcare, transportation;

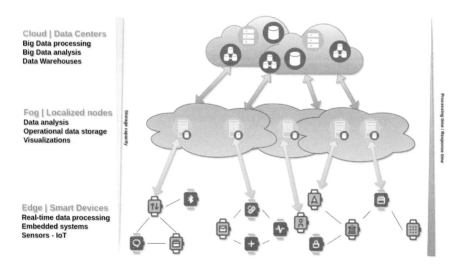

Figure 6.2: Big data environment for time series storing and analysis.

- industrial applications, such as, agriculture and manufacturing;

- infrastructure applications, for example, environment monitoring, energy management, water management;

- military applications.

Applications at the Fog Computing layer can store and manage operational data, that is, data that is recent, using more complex Database Management Systems, for example, InfluxDB or Kdb+. At this layer, analysis can be done using frameworks that provide less computational intensive Machine Learning algorithms to construct models for forecasting, anomaly detection, and change point detection. These models are always trained on recent data to extract patterns and provide new knowledge for the short term decision-making process.

Applications at the Cloud Computing layer can provide data storage capabilities for historical data, using complex database systems such as TimescaleDB and OpenTSDB. Complex models for data analysis can be employed at this layer, such as Machine Learning algorithms and Deep Learning architectures that require computational power for intensive calculations and increased resources for processing in-memory large Time Series data.

6.6 Conclusions

In this chapter, we have provided an overview of all the steps required for storing and analyzing raw Time Series data. First, we have formalized Time Series mathematically and then provided some insights on Time Series forecasting, outlier detection, and change point detection. Second, we discussed the NoSQL concepts and data models by critically presenting the advantages and disadvantages of NoSQL database systems. Third, we presented the architectures and data models of the main NoSQL Time Series databases. We compared the principal characteristics that are required for deciding on the storage system to use for applications. Finally, we presented the Big Data environment required for deploying storage and analysis applications for handling Time Series data.

Time Series Database Management Systems prove to be a good choice for storing data collected from the Internet of Things. Depending on the use case, the architectural design can provide some facilities such as horizontal scale, concurrency control, replication, reliability, and storage type. The query language and its operations can also play an important factor in choosing one system over the other. Some language functionalities that need to be considered are related to the ability to use CRUD such as, filtering, aggregation, and JOIN.

Raw Time Series data can be collected from the Internet of Things sensors and smart devices and then sent to the Fog Computing layer. At this layer, data can be analyzed promptly by pre-trained models to predict future values and detect anomalies and change points. The information can be stored using simple data management systems. Alerts can be issued in real-time to inform users about changing values and parameters in the monitored systems.

From the Edge Computing layer, the raw Time Series can be sent to the Fog Computing layer where it is stored in the operational Time Series Database Management Systems. At this layer, analysis can be more complex and can help the short term decision-making process. Users can make informed decisions about the current status of the monitored systems and devise business strategies in accordance. Forecasting using regression models or shallow neural networks architectures, outlier detection using clustering techniques, and change point detection using approximate search methods can be employed at the Fog Computing layer.

From the Fog Computing layer, data can be sent for long time storage at the Cloud Computing layer, where it will be stored in dedicated Time Series Data Warehouses. At this layer, the analysis algorithms can be more complex and multi-faceted. Computational intensive Machine Leaning algorithms and Deep Learning architectures should be trained at this layer. The available computational power provided at this layer can be used to do extensive multi-dimensional forecasting and outlier detection using deep neural network architectures. The models can be fine-tuned using hyperparameter tuning. The historical data can improve the accuracy of the machine learning algorithms. Business intelligence analysis and optimal solution algorithms for change point detection can be performed at this layer.

Chapter 7

A Territorial Intelligence-based Approach for Smart Emergency Planning

Monica Sebillo,[a],* *Giuliana Vitiello,*[a] *Michele Grimaldi*[a] and *Davide De Chiara*[b]

7.1 Introduction

The term Spatial Data Science gathers together traditional spatial data analysis, geo-processing and big data analysis, including machine learning and deep learning techniques [130], [184], [196], [21], [231], [71], [214] to uncover hidden patterns and improve predictive modeling. These disciplines can contribute to the development of geospatial big data analysis and processing and can be used alone or together with other types of big data to acquire knowledge from geospatial big data. Once such knowledge has been produced, it can be integrated and elaborated by business intelligence processes [69], thus providing users with tools for territorial intelligence. In particular, the concept of Territorial Intelligence (TI), also known as geographic intelligence, is gaining an important role for organizations and companies belonging to the same geographic area because

[a] Dipartimento di Informatica, Università di Salerno.

[b] Italian National Agency for New Technologies, Energy and Sustainable Economic Development (ENEA).

* Corresponding author: msebillo@unisa.it

it works as a collector of diverse knowledge and contributes to improve the exchange of strategic information at the local level [211], [173], [99].

This form of knowledge, accessible and deployable by general-purpose users, represents the first step towards a systemic approach to build networks and clusters of domains, such as environment and infrastructures, on which specific actions can be applied, thus enhancing their effects. On this basis, the ICT support is paramount for contributing to sustainable development by expanding the impact of TI processes in a broader context. Figure 7.1 describes relationships among actors, namely functionality and resources aimed at making it a territory which could finally take advantage of collective knowledge, innovate, and create cross-fertilization to benefit different domains.

The above relationships are strongly beneficial because they can play a significant role within several activities having the involvement of local stakeholders and citizens, who can express their needs and share their knowledge, which is desirable.

In [320], a scenario as described, the involvement of expert and ultimate users contributes to the improvement of the territory, in terms of niche tourism offers and well-being of diabetes patients. The paper shows how TI can create connection among apparently different fields and exploit territorial peculiarities emphasized by the knowledge discovery processes. This goal has been achieved through an effective ICT- and TI-based support to a complex spatial decision making problem, which requires techniques and tools that help articulate a decision goal, discerns information from the multitude of data through data analysis and visualization, creates plausible scenarios representing possible courses of action, computes and visualizes their impacts, tests the stability of scenarios, and prioritizes them to help select the right course of action.

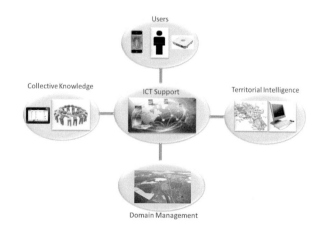

Figure 7.1: TI-based management for spatially-enabled territory and communities.

The goal of this chapter is to extend the applicability of such an approach and show how TI processes can create valuable knowledge about city dynamics to support territorial governance in decision making processes. In particular, the focus is put on the area of informational urbanism, where the inclusion of new concepts, such as big data, data analytics and TI, produces an unlimited potential development capable of meeting principles of sustainability, quality, equity and rapidity, the main pillars for smart city planning.

The chapter is structured in two parts. The first aims to provide an overview of existing methods and technologies for managing big data that have found a natural integration in spatial data science. The goal is to highlight how this discipline brings everything related to GIS and Analytics in a single context to provide important support for the technological transformation taking place in different sectors, such as Smart Planning.

The second part of the chapter addresses the role that TI can play in bringing innovative results for the management of spatially-enabled territories and communities. In order to demonstrate the feasibility of such an approach, Smart Planning is recalled and used as an appropriate method to guarantee the updatable and integrable feature of the Emergency Planning domain, where a large amount of data has to be collected to derive territorial knowledge useful for decision making. In particular, the risk management, a relevant aspect of this governance, represents a context where experimenting methods for big data integration and insights discovery is fundamental to achieving successful approaches for a rapid and efficient emergency decision making process.

7.2 From GeoSpatial big data to territorial intelligence

A large part of all big data, as anticipated, has a spatial component, that can be associated with a specific set of coordinates through a process known as geocoding, which translates addresses and points of interest into geocodes. Research in (geospatial) big data domains offers several multipurpose methods and techniques, known as Advanced Analytics (such as, predictive analysis, data/text mining, machine learning, sentiment analysis, neural networks), all aimed at better understanding trends and getting insights from large amounts of data. The goal of this section is to provide readers with an overview of the basic concepts of this domain.

7.2.1 Geospatial big data

The concept of big data is defined through three statements, namely huge volumes of data, complexity of data types and structures, and speed of new data creation and growth. Big data is also associated with a "3V's" framework for its understanding and management, initially introduced by Laney. In [208], the author proposes three dimensions that characterize the challenges and opportunities of big data: Volume, Velocity and Variety (3Vs). Then, the framework was enriched to better support early detection of big data characteristics for its classification, and the Veracity and Variability parameters have been added to define the 5Vs Big Data framework. This widely accepted characterization method also applies to geospatial data, which has at least one of the 3Vs [115], such as the variety of data types in spatial computing and the computational expensiveness of spatial analytics, although the other Vs may be relevant, such as Visualization and Visibility. Indeed, with a conspicuous availability of new types of sensors and communication technologies, new ways of collecting geospatial data have materialized, leading to completely new data sources and data types of geographical nature [103]. In particular, geospatial data can be collected by using ground surveying, photogrammetry and remote sensing, and through laser scanning, mobile mapping, geo-located sensors, geo-tagged web contents, volunteered geographic information (VGI), and global navigation satellite system (GNSS) tracking. The feature shared by all these types of data sources is that these new data types extend and enrich geographic data in terms of thematic variation and the data itself is more "user-centric". The latter is mainly true for VGI that come from social media [125].

Once data collection capacity is no longer hampered by progress and innovation, the most important question is how to exploit this amount of geospatial big data [215]. The above general characteristics imply that big data cannot be easily caught, stored, analyzed and managed by conventional hardware, software and database tools and requires new technical architectures and analytics for producing business value [243]. Toward this direction, many efforts have been made and various infrastructures have been designed and made available to transform data into useful information. Furthermore, as big data has become a strong focus of global interest that is attracting more and more attention of academia, industry, government and other organizations, the study of geospatial big data can be placed in the disciplinary area of traditional geospatial data handling theory and methods. However, the increasing volume and varying format of collected geospatial big data imply major challenges in many phases of its life cycle, mainly in storing and analyzing them, due to their impact on the resulting data quality.

In general, systems for collecting and analyzing geospatial big data consist of three layers: geospatial big data integration & management, geospatial big data analytics, and geospatial big data service platforms. While the third layer heavily depends on third-party providers that deliver hardware and software tools to users over the Internet as a cloud computing model, the first and the second layers evolve based on the research efforts made in the corresponding fields. In particular, the first layer is responsible for quickly storing, retrieving, indexing, and searching geospatial big data. To this end, the main challenge is both to organize geospatial data through spatial data clustering and to design spatial indexing methods to make query processing faster [369], [370], [371]. The second layer, often the most complex, is responsible for performing data analytics. It is generally structured as a module for interactive analysis of real-time or dynamic data and a module for batch analysis of static or archived data. The following Subsections 7.2.2 and 7.2.3 present an overview of analytics and visual analytics applied to geospatial data.

7.2.2   *Spatial data analytics*

Spatial Data Analytics focuses on algorithms whose goal is to uncover actionable insights for current problems, thus leading to immediate improvements. In particular, with questions in mind that need answers ("they know or they don't know the answer to"), spatial data analysts aim to solve problems by processing and performing statistical analysis on existing spatial datasets.

Most well-established architectures for spatial data analysis incorporate a spatial On-Line Analytical Processing (SOLAP) engine which aims to combine Geographic Information Systems (GIS) and OLAP [189]. It is a visual platform for analyzing spatio-temporal data through a multidimensional modality with aggregation levels being available on cartographic displays and in tabular and diagram displays. The engine can perform analysis by conducting several operations such as roll-up, drill-down, and aggregation on the geospatial data warehouse. The result is displayed on various types of maps, so that a user can easily grasp the result of each operation.

Geo-Mondrian is the first implementation of a true SOLAP server [161]. Developed as a spatially-enabled version of Pentaho Analysis Services (Mondrian), it is an Open Source SOLAP Server that allows users to embed spatial analysis capabilities into analytical queries, thus producing true geo-analytical queries and a real integration of spatial objects into the OLAP data cube structure, instead of fetching them from a separate spatial database, a web service or a GIS file.

While SOLAP represents a general-purpose approach to spatial analytics, recent solutions are aimed at benefiting from advances made by big data-oriented methods and technologies. Some new approaches are in fact designed to extend Hadoop to geospatial big data and therefore to utilize its spatial version as a base platform, that is Spatial-Hadoop [105], [106]. In particular, there is a growing interest in using the MapReduce framework, the Hadoop component for data processing, to perform parallel processing of huge volumes of data by dividing tasks into a set of independent activities. Once each part has been processed, the partial results are properly assembled into a unique final result.

7.2.3   GeoVisual analytics

In [351], Thomas and Cook define visual analytics (VA) as the science of analytical reasoning facilitated by interactive visual interfaces. In [198], Keim et al. provide a more detailed definition where visual analytics represents a combination of automated analysis techniques with interactive visualizations, coming from the Information Visualization (IV) discipline, for an effective understanding, reasoning and decision making based on very large and complex data sets.

When applied to geospatial data, VA can benefit, in a single context, from both the current abundance of geospatial data and tools from GIS and IV disciplines. GeoVisual Analytics (GVA), in fact, puts its focus on analytical reasoning rather than visualization itself, to help people analyze and make sense of geographic information through visual interfaces, by finding patterns within that information [297].

In [236], MacEachren gives the following definition of GVA, "Geovisual Analytics is a domain of research and practice focused on visual interfaces for analytical methods that support reasoning with and about big, dynamic, heterogeneous, unconfirmed, hyper-connected, geo-information – to enable insights and decisions about something for which place matters." This definition emphasizes the three main features of GVA, namely an activity (analytical reasoning), a technology (interactive visual interfaces) and complex data. Hence, the goal of GVA is to derive knowledge from data by combining visual data exploration and computational processing [317].

Besides the role of visual interfaces, MacEachren emphasizes the complexity of data on which analytical methods are usually applied. The discipline of GVA has affected, in fact, several domains in different ways, until 2007, when Andrienko et al. in [20] set the research agenda. They stimulate concerted efforts to tackle space-related problem solving and build a research community that gathers the appropriate competences in that domain. In particular, they suggest the name "Geovisual Analytics for Spatial Decision Support" for the cross-disciplinary research conceived to improve human capabilities to analyze and reason about space-related decision problems, starting from the more general research discipline of VA [352]. From VA, in fact, some key points are imported and, when necessary, adapted to the geospatial data complexity, such as the need to create tools and techniques to enable users to synthesize information and derive insight from big data, detect the expected and discover the unexpected, and provide and communicate timely and understandable assessments.

Since then, much work has been done following the indications proposed by the agenda, which suggests classifying problem types according to some candidates, such as spatial extent of territory, temporal extent, domain, complexity and number of decision makers.

7.2.4 Geospatial business intelligence and spatial data science

By discussing the state of practices in data analytics from a different perspective, a further classification is possible based on the issue under investigation. Four types of big data analytics can be formulated, as follows:

1. Descriptive analysis, it allows users to answer "What happened?" questions;

2. Diagnostic analysis, it allows users to answer "Why did it happen?" questions;

3. Predictive analysis, it allows users to answer "What could happen in the future?" questions;

4. Prescriptive analysis, it allows users to answer "How should we respond to those potential future events?" questions.

These analytical techniques emphasize the different role that Business Intelligence (BI) and Data Science (DS) can play. BI aims to explain current or past behavior by aggregating and grouping historical data. It produces reports, dashboards and queries based on structured data and traditional sources. Common questions are "when" and "where" events occurred. By comparison, DS, or data-driven science, applies an exploratory approach focused on insight about current activities and foresight about future events. By using predictive modelling applied on very large (and unconventional) datasets, DS tries to answer questions related to "how" and "why" events occur. In particular, DS, which emerged within the field of Data Management, uses machine learning and artificial intelligence techniques to extract meaningful information and to predict future patterns and behaviors, both tasks usually requiring human intelligence. Thus, business users can translate knowledge into tangible business value, that is, they can add a new V (value) to the 5Vs framework. The previous classification still holds when BI and DS are applied to geospatial big data. In both cases, the goal is to provide expert users with algorithms to uncover hidden patterns and improve predictive modelling when data has a spatial component and place-based context is the keystone to achieve this goal.

7.2.4.1 Geospatial business intelligence

Geospatial Business Intelligence (GeoBI) is business intelligence that makes use of geospatial information. In general, it combines GIS and BI technologies to better support the data analysis process and help users make more efficient decisions [274]. In particular, GeoBI offers solutions to identify both schemas underlying relevant trends and correlations among spatial parameters within a complex (un)structured, historical/current dataset. The growing interest in GeoBI stems from its huge potential as an enabler for many mass-market geospatial-centric applications, such as location based services (LBS) and Web mapping.

7.2.4.2 Spatial data science

Spatial Data Science (SDS) [22], [102] incorporates tools from multiple disciplines to collect and process spatial data, extract meaningful knowledge from large amounts of data and derive insights, and interpret it for decision-making purposes. Disciplinary areas that are involved in the SDS field include mining, statistics, machine learning, analytics, and programming. In particular, machine learning plays a paramount role because it finalizes the decision model derived from predictive analytics by matching the likelihood that an event occurs to what actually happened at a predicted time.

Spatial Data Scientist's goal is to pose questions about spatial phenomena and scenarios and identify potentially new lines of research. Less concern is devoted to finding the right answers, while attention is paid to the formulation of the right question to find solutions to problems that have not yet been thought of, that is, to stimulate the answers to things "you don't know." A further difference between SDS and Geospatial Big Data Analytics [217], which sometimes are used interchangeably, is the scope. SDS is a comprehensive concept that includes other fields to mine large datasets. Also, while Geospatial Big Data Analytics works well when focused on specific questions that need to be answered, SDS concentrates on which questions should be posed, that is, SDS is focused more on asking questions than finding specific answers.

7.2.5 Territorial intelligence

GIS without intelligence have only a little chance to provide effective and efficient solutions to spatial decision making problems [124]. Generally speaking, TI is a system of models, methods, processes, people and tools that allows for a regular and organized collection of data generated by territory. Through processing, analyses and aggregations of such a data, TI allows its transformation into information, its preservation and availability, and its presentation into a simple, flexible and effective form to constitute a support for strategic, tactical and operating decisions [320].

In literature, there exist different definitions of TI, each of them focusing on a specific aspect of the discipline. Initially, the concept of TI had been associated with the sustainable development goals, as a means to foster the interaction among its three pillars, namely economy, society and environment [211]. Much work has been done along that line, where the specific role of TI was to integrate computer intelligence with human collective intelligence to achieve a sustainable development for any territory through its local administrations and local politicians [91]. In [136] the authors define TI as the science with an ability to connect the sustainable development of territories and territorial communities by a multi-disciplinary knowledge. In [50] Bertacchini refers to TI as the sharing of know-how about the resources of a territory, between categories of local actors of different cultures. Miedes Ugarte [253] splits TI into three main components, namely cognitive, socio-political and organizational-technological, and puts the emphasis on the last one, which is addressed to the analysis, monitoring and territorial communication. In [210], the author defines TI as the composition of collective human intelligence allied to artificial intelligence for sustainable development (Fig. 7.2).

CAENTI (Coordination Action of the European Network of Territorial Intelligence) is an interesting project funded under FP6-CITIZENS whose goal was to integrate coeval research projects on TI tools, so as to give them a European dimension [83]. Project activities focused on designing tools and scientific methods useful for territorial information analysis and the involved partners payed close attention that the tools were complied with principles of good governance and ethics of sustainable development.

The goal of the following sections is to show how the applicability of TI processes can be extended to support territorial governance in decision making by fostering a participatory urban planning. The rationale under this approach is the awareness that Location is a unifying element for territorial services oriented to citizens. Location can be an address, a route, or an area, and spatial relationships, patterns and trends that can be derived from large amounts of this element can bring invaluable benefits to business applications. Hence, more and more public institutions and private industries are adopting the approach of integrating human knowledge and machine-processable knowledge, derived from spatial analysis and BI, to obtain the basis for TI.

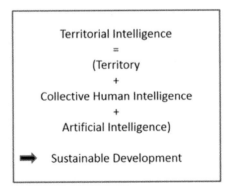

Figure 7.2: Definition of TI according to [210].

7.3 Smart planning for risk management

The concept of smart planning emerges as an alternative to traditional planning. It is based on the integration of a large amount of data and information and supports the decision-making process quickly and responsively. Moreover, it defines a new form of urban plan, named smart city plan, which is founded on both people's engagement and usage of innovative technologies as a support for land management.

The use of a smart planning approach is particularly useful in assessing risk and vulnerability. In this domain, in fact, many factors and variables related to the occurrence of catastrophic events could not have reached a level of providing evidence that makes them explicit. This situation mainly occurs in highly dynamic territorial contexts where both vulnerability and socio-natural factors could undergo rapid changes in the short and medium-term [131].

A further challenge for planning methodologies derives from the importance of integrating uncertainty into risk management. The need to clarify this topic assumes particular importance in redefining both the contents of the Municipal Emergency Plan (MEP) and the planning tools used for this activity. In particular, emergency planning requires a revision of its operational tools through new performance-based approaches, capable of analyzing the complexity of the existing spatial relationships. Infact, a PEC must provide an objective and numerical assessment of critical territorial issues relating to the investigated process, and ensure an effective response in terms of risk and resource management.

The management model proposed by the civil protection manuals is often based on specific knowledge of the territorial risk status, which is derived from old databases. This jeopardizes the character of the PEC, which, unlike other planning tools, must instead be updateable and integrable.

With this regard, Smart City Planning could play an important role in taking into account capabilities, such as data collection, information processing, real-time monitoring and networked infrastructure management, useful to the achievement of this goal.

In order to support all these activities, a Spatial Decision Support System (SDSS) has been developed capable of automating the process of drawing up a PEC and producing assessments highly detailed on the one hand, and easily accessible on the other. The methodology proposed for the SDSS construction involves three macro-phases. It is based on the definition of territorial risk, intrinsically linked to the possibility that any phenomenon (natural or induced by human activity) may generate harmful effects on population, settlements and infrastructures, within a geographical area and in a given period. The variables that contribute to its definition are numerous and complex. According to UNDRO (United Nations Organization for Disasters), they can be always grouped into three macro factors,

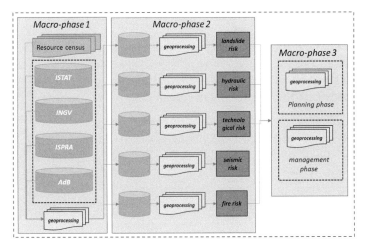

Figure 7.3: Workflow of the methodology.

suitably combined, namely Hazard, that is the probability that a phenomenon of a given intensity occurs in a given period and in a specific area, Vulnerability, i.e., the propensity of an element (people, buildings, infrastructures) to suffer damages caused by the event stress of a certain intensity, and Exposure, which corresponds to the number of elements at risk present in the considered area, such as human lives, settlements and infrastructures.

Figure 7.3 details the three macro-phases. The first macro-phase proposes data acquisition and storage through an intelligent census based on the Web of data approach [162]. In the second macro-phase, geoprocessing models are used to automate spatial analysis and data management processes to derive territorial knowledge supporting the emergency plan. The third macro-phase proposes a model for the automation of procedures necessary to the emergency phase management. The procedure follows the object-oriented design approach and is based on a set of sequence diagrams that illustrate the order in which the domain objects, already modelled in a class diagram, interact with each other [55].

7.4   A spatial decision support system for risk management

Once the SDSS has been implemented, it allows planners to quickly estimate the risk indices resulting from the available information. More details are given in the following subsections, where the system is illustrated when applied for the emergency planning.

7.4.1 The emergency planning

The current emergency planning is aimed at providing the best rescue through a preventive solution to problems that the rescue bodies face at the time of the event. This activity takes place by both a drafting of the PEC and the Emergency Limit Condition (CLE) for the urban system [93].

A PEC represents the means allowing authorities to prepare and coordinate rescue interventions addressed to protect population and property in a given risk area, maintaining a level of civilian life anyway, put in crisis by a situation involving physical and psychological damage. A PEC has a well-defined structure consisting of three parts, namely a General part, which collects information on territorial characteristics and structure, a Planning outline, aimed at describing risk scenarios, and an Intervention model, conceived as a set of choices and procedures addressed to overcoming the occurring emergency. A CLE is used to determine the minimum conditions for overcoming the emergency following a seismic event. In that case, the interruption of all residential, urban and strategic functions is accepted, while the conservation of the emergency management functions is necessarily required.

In order to guarantee the correct execution of the risk management activities as planned for both a PEC and a CLE, a periodic update of territorial knowledge has to be carried out, which takes into account the evolution of the territorial structure and changes in the envisaged scenarios. Numerous essential variables are then investigated for an accurate and upgradable plan, which refer to several scenarios, each corresponding to a given risk event, multiplied by the number of exposed areas and amplified by different boundary conditions. In particular, defining a damage scenario results in the appropriate tool to achieve the right level of territorial knowledge. It allows the definition of a territorial framework of the area involved in the event, then providing relevant information, such as the location and the extent of the most affected area, the functionality of both transport networks and communication and distribution lines, the expected losses in terms of human lives, injured, homeless, collapsed and damaged buildings, and the corresponding economic damage, with an obvious impact on Civil Protection activities, both in the emergency planning and management activities. This information also supports the identification and description of the occurred event(s), in order to size human resources, materials to be used and their allocation, as required in the plan, thus immediately providing a description of the real event and its impact on the territory.

A combination of all aforementioned activities has an implicit complexity, being the territory a domain that evolves with spatial and temporal continuity. Such a difficulty in the acquisition, management and updating of the parameters appears to be a significant burden that most municipalities are unable to comply with, due to the scarcity of resources and lack of qualified personnel, even if the procedure is well codified by the civil protection manuals [262].

7.4.2 *A SDSS for the emergency planning*

To develop the proposed SDSS, three components have been designed by applying the three macro-phases described in Section 3, which are in turn consistent with the three parts of a PEC procedure. The approach is mainly focused on data collection, thus guaranteeing a considerable simplification of the procedure itself. In fact, the proposed model automatically elaborates each required information on the basis of territorial data provided. Data sharing and components interoperability represent the two key properties that reduce data inconsistency, errors and useless repetition of data processing, which may cause a relevant increase of processing time. The three macro-phases of the model can be treated independently in three separate toolboxes, each containing models capable of meeting the requirements laid down in the manual for drafting/updating a PEC. To create geoprocessing models, a graphic modeller is used. The resulting graphic model is represented as a diagram that chains together sequences of processes and geoprocessing tools, using the output of one process as the input for another process. It handles variables as parameters and establishes workflows that can be shared among users. The use of a graphic model is necessary due to the amount of data required by the operating manual, which, for the PEC drafting, requires a series of geometric and analysis operations which are repetitive, univocal and standardized, independently of the context and territory of analysis/study. The outcome of the graphic model represents the complex articulation well inherent in drafting a PEC. In addition to data retrieval, the workflow globally implements 324 spatial analysis functions and 464 variables. To test the flexibility of the resulting model, parameterization of the variables was used for, calibration by application to different municipal areas in southern Italy, such as Montetorte Irpino (AV), Teano (CE), Barano d'Ischia (NA) and Ponte BN. The following example describes some activities carried out while practicing the complete model to the Municipality of Chiusano San Domenico (AV) in the South of Italy.

7.4.2.1 *Macro-phase 1*

The macro-phase 1 is embedded into a partially sequential process that, based on a resource census, builds a sharable data repository, and aims to produce territorial knowledge useful to the risk management activities.

The whole process starts by acquiring and storing data through an intelligent census based on the Web of data approach [151]. An ontology of the domain is under construction using the Protégé ontology editor [262], [142], which allows modelling and validating resources in terms of metadata, state and behavior [319]. Moreover, the ontology can be used to aggregate subsets of resources to build units to which a composite state and a behavior can be associated. Inference algorithms can obtain additional relationships among validated resources.

The goal is to have a system that dynamically generates new cards for data entry tasks and populates the available dataset through procedures which are based on effective territorial knowledge, where instrumental and human resources, rules, behaviors and interactions converge. A set of main classes have been identified, covering different areas (for example, attending and recovery areas), alert methods, buildings, streets, vehicles and equipment, that can be used in the definition of emergency plans. To this aim, the design of data started from a set of 97 cards (each card had been created by taking the identified cases into account). Those cards along with the guidelines to perform the analysis of CLE represented the basis for identifying the digital census metadata (Fig. 7.4).

A fundamental feature of the system, whose results are simple and intuitive to use, is the "dynamism of data", meant to constantly update resources and their possible interactions. A high-quality standard also has to be guaranteed in order to avoid duplications, inconsistencies, incompleteness and unreliability. Figure 7.5 shows the interface available for data entry activities. It classifies data in terms of Plan, Matrix, Theme, Class and Resource.

Figure 7.6 illustrates a relevant step of data acquisition, namely the association between a resource and its spatial location expressed as a vector geometry [318]. To this goal, Google Maps API have been used, the Drawing Tools in particular that exploit the Drawing Manager class, used to select and assign a card a physical area.

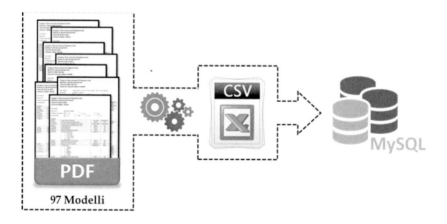

Figure 7.4: Process for metadata and data definition.

Figure 7.5: The initial selection for data classification.

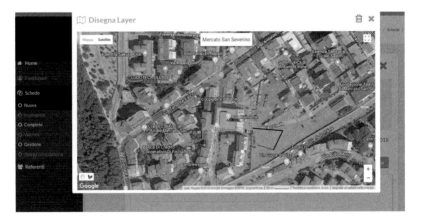

Figure 7.6: The interface for drawing a geometry.

7.4.2.2   Macro-phase 2

As for the second macro-phase, a model for data recovery is generated (Fig. 7.7). It accesses the more general databases made available by regions, provinces, basin authorities and municipal offices and extracts data related to the area of interest. The construction of risk scenarios, have been obtained by an explicitly spatial combination of elements exposed to hazard and vulnerability.

The graphic models that refer to different risk scenarios are shown in order of complexity. This is essential both to understand how processing can be simplified in terms of time and guarantee the process control by the operator thanks to the complete parameterization of models. As for the hydrogeological risk scenario

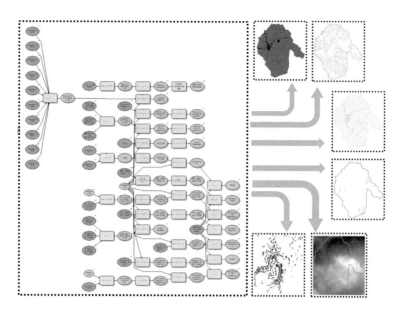

Figure 7.7: The graphic model for data recovery.

and the technological disaster risk scenario [112], the corresponding two models are based on a spatial overlay, that is a selection of the exposed elements in terms of buildings and people with respect to the areas of high and very high hazard (Fig. 7.8). In particular, for the scenario of technological disaster risk, the model performs a selection of the activities in the involved area, considering those present in the inventory of the Ministry of Environment, and subsequently generates the hazard bands through buffering operations.

In relation to the seismic scenario, given the usual lack of precise knowledge about the structural characteristics of buildings, good practices provide for the use of a simplified model to estimate the number of displaced and homeless (Fig. 7.9). It is based on the ISTAT census data that refers to the age of constructions [133], types of masonry and geological conditions taken from seismic micro-zonation maps.

As for the construction of the fire risk scenario, the procedure becomes much more complex since it must respond to a set of complex criteria to be spatially modelled. Briefly, the first operation involves the generation of the interface, that is the area in which the interconnection between anthropic structures and natural areas is very narrow. It is subsequently classified on the basis of 6 criteria that influence the progress of a fire, namely the vegetative typology, its density, the slope of the area, the type of contact, the proximity to previous fires, and the sensitivity of the exposed area (Fig. 7.10).

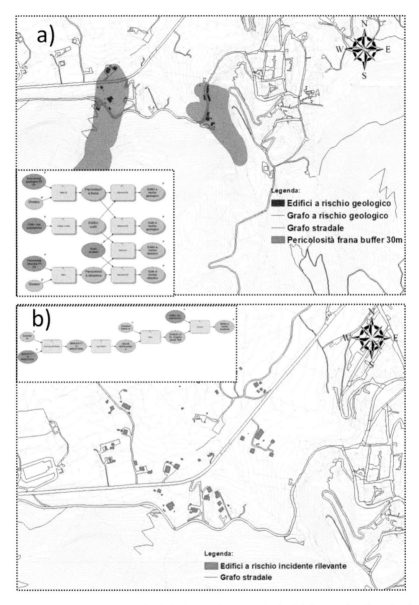

Figure 7.8: The Graphic model for building: (a) hydrogeological risk scenario. (b) technological disaster risk scenario.

7.4.2.3 Macro-Phase 3

The Intervention Model translates the actions to be taken in response to the activation of an event scenario in terms of operating procedures and protocols. It

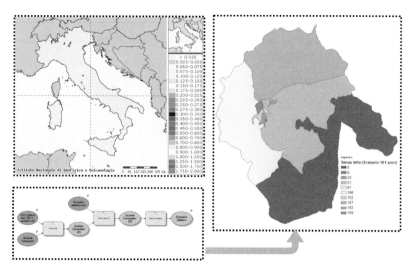

Figure 7.9: The graphic model for building the seismic risk scenario.

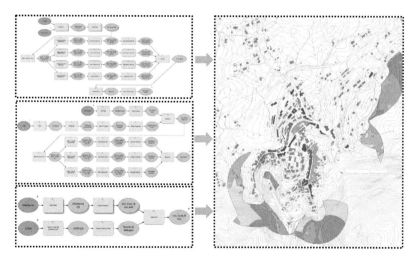

Figure 7.10: The graphic model for building the fire risk scenario.

includes a planning and a management phase. As for the planning phase, the intervention model establishes the identification of the emergency areas starting from a datasheet, which allows the selection of the lots that comply with some necessary conditions expressed as a set of criteria. Once the main elements of territorial knowledge and risk scenarios have been defined, the subsequent activity is aimed at both identifying and verifying the emergency areas, and analyzing the CLE. This phase responds to a set of criteria coded by the Civil Protection.

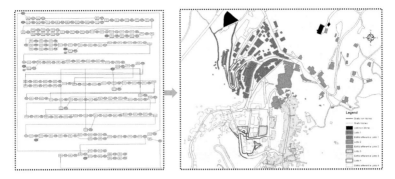

Figure 7.11: The graphic model for identification of the area of influence of each waiting area.

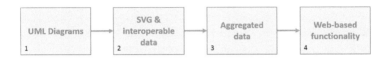

Figure 7.12: The process for metadata and data definition.

In the following step, the model assigns each building the closest suitable area. In case of buildings located at a distance greater than 300 m from the nearest emergency area, a new waiting area should be identified on site. The model automatically checks the size of population the lot can host. Moreover, it supports both the compilation of CLE sheets and the selection of emergency areas and structural aggregates. As a matter of fact, it assesses the characteristics of the interference of buildings with emergency areas and safety routes. Figure 7.11 shows the model output, namely the suitability of the escape paths, the identified portions of influence of each waiting area and the identification of unsuitable areas with respect to criteria established by the PC manual.

As for the management phase, a prototype has been developed through which it is possible to connect the guidelines modelled as UML diagrams [284] and the validated instances of the available resources. Figure 7.12 shows the overview of the process, starting from the census previously described, built territorial knowledge useful to the risk management activities.

In the first step of this phase, the guidelines of the Augustus method are expressed as UML class and interaction diagrams [135]. The former characterizes resources and actors involved within a plan; the latter describes the interactions among them during the diverse operation phases, namely Attention, Alert and Alarm. By interacting with the interface of the UML diagram editor, the rules of the method can be easily modified and the updating is automatically applied to the whole process.

The second step translates those diagrams into SVG files which are then parsed to extract information about classes, relations, interactions and activities. These two phases express the rules as a series of elementary and interoperable data representing the relevant concepts of the domain.

The development of an emergency plan ends with a validation phase aiming at facing possible exceptions caused by both human factors and temporary objective impediments, such as a work in progress on a road network. During that phase targeted training activities are scheduled, which may contribute to tune the involved parameters (residents, personnel and tools) of the underlying protocol, by taking into account both general requirements set by national regulations and local availability and supply. Results consist of modifications and instructions to be integrated within the initial intervention model.

The third step is addressed to aggregate the above information according to precise data structures elaborated to then be easily used in various functionalities. At this stage, a fundamental requirement is the interoperability, since it allows integrating data coming from any other parsed source. Figure 7.13(a) shows the interface by which SVG files can be downloaded, whilst Fig. 7.13(b) shows a selected class and classes to which it is related. Finally, Fig. 7.13(c) displays the activities involving the selected class.

The selected class can be also highlighted both in the Class Diagram (Fig. 7.14(a)) and in the Interaction Diagram (Fig. 7.14(b)) interface.

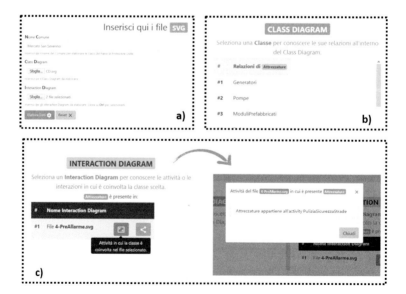

Figure 7.13: The interface to download SVG files (a). Classes related to the selected one (b). Interactions referring to the selected class (c).

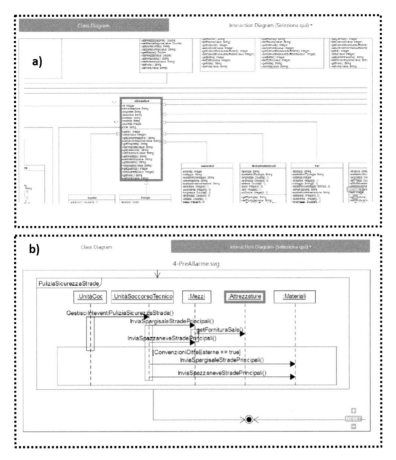

Figure 7.14: Interacting with the class diagram (a). Interacting with the interaction diagram (b).

Once the census has been completed, it is possible to link the general-purpose result of this process to the instances of validated available resources. This functionality is fundamental to guarantee that the adopted emergency plan is always updated.

7.5   Conclusion

A smart city is more than a high-tech city and people are the real protagonists in these changing city scenarios. Thanks to the emerging technologies, citizens are in fact important "data sources" that help the development of citizen-centered smart services in several domains, such as urban planning and mobility. In these fields, knowledge is built when a person or an application translate information into actionable insights for decisions making, that is, when information is put into practice.

The goal of this chapter was to describe a territorial intelligence based approach conceived to support smart planning activities, where the role of different users is specified in agreement with expected functionalities. The research results materialized in a SDSS, tested in different municipalities.

The innovative aspect of the underlying methodology is the possibility of being incorporated into the emergency operational plans, offering planners, local authorities and stakeholders a useful decision support tool.

The application of the model showed the robustness of results and the effective possibility of improving the emergency plans in the direction of developing a SSD system capable of real time processing of the risk conditions when activated by a given scenario. SSD can immediately provide the involved population with a set of actions to be carried out (such as, lots people having to reach and roads they have to take), depending on the particular risk condition activated, not just as a function of an a priori prediction. In particular, the application of the model shows the possibility of intervention on two needs that substantially affect the PEC and CLE contents and the related intervention model. As for the first case, the requirement is to update the contents according to the evolution of the urban system, which changes over time in terms of land use, age of the built heritage and demographic and location dynamics. That is, the risk scenarios, the CLE and, consequently, the review of the emergency areas must be periodically updated.

As for the second case, the requirement is to optimize the cognitive frameworks supporting the intervention model. As a matter of fact, during the pre-alarm and alarm phases, it is necessary to re-elaborate the risk scenarios according to both the real distribution of the settlement loads (in terms of population) and the modification of escape paths. In both cases, the model allows processing the required information, excluding the acquisition and updating of the knowledge framework.

Outcomes obtained prove that better results can be achieved when common issues are jointly tackled, since the integration of results represents the first step towards improved geospatial big data analysis applications, especially when the volume and the complexity of data is an obstacle to reach an efficient solution.

Chapter 8

Big Data Analysis and Applications for Energy Performant Buildings and Smart Cities

*Ezilda Costanzo** and *Bruno Baldissar*

8.1 Introduction

Buildings are responsible for 40% of the EU's total energy consumption and 36% of CO2 emissions in the EU, making them one of the biggest contributors to global warming emissions in Europe.

MS have set ambitious targets for their building stock by 2050: increased electrification can potentially make integration of renewables, energy storage and demand-side management more effective. Business opportunities for utilities and energy services companies involved in building-related energy efficiency technologies like Building Energy Management (BEMS) are rapidly increasing.

Nearly zero energy buildings is one of the targets for a low-carbon economy in 2050, but new construction rates at present are very low and the existing stock is obsolete and low performing. Better evidence-based decisions are needed to

Italian National Agency for New Technologies, Energy and Sustainable Economic Development (ENEA).

* Corresponding author: ezilda.costanzo@enea.it

improve performances of the building stock in the wider energy systems. Nevertheless, energy data in buildings is still too fragmented and lacks interoperability.

In order to support wide-scale data collection and processing, the European Commission is nowadays calling for "national, regional and local authorities, property management companies, technology providers and stakeholder associations from relevant sectors to stimulate and enable a comprehensive and long-lasting community committed to improve and strengthen data collection"[1].

Several EU projects[2] have already collected relevant information. The Building Stock Observatory (BSO), launched by the European Commission in November 2016 as part of the "Clean Energy for All Europeans" package to monitor the energy performance of buildings in the EU, developed a methodology for data collection and a website. So far, only 13% of the 250 indicators per MS, that include, beyond building physics, energy poverty, embodied energy, indoor and comfort data, have been populated in the BSO database [114].

Yet, routine assessment, data consistency and quality, terminology and definitions are still the main concern in these initiatives. The convergence of big data, block chain and the Internet of Things (IoT) is of particular relevance to improve performance at a single building, district and city levels. Analysts are just starting integrating building data coming from sensors and meters into smart cities planning[3]. This paragraph illustrates experiences in the EU to encourage/support collection of data towards big data applications in the building and urban context, with a focus on:

- ■ Which data is being collected or integrated

- ■ Stakeholders engaged, target groups and users

- ■ Which services can be provided

- ■ Barriers on data formatting and interoperability towards more dynamic and automated collection ("live picture" of performance).

8.2 Which data are collected and integrated to guide decision making

Building energy data may be generated from design and construction (for example through BIM, Building information modelling), and post occupancy evalua-

---

[1]H2020 LC-SC3-B4E-7-2020 Call, "European building stock data 4.0" https://ec.europa.eu/info/funding-tenders/opportunities/portal/screen/opportunities/topic-details/lc-sc3-b4e-7-2020.

[2]For example, the *EUEPISCOPE, ZEBRA, ENTRANZE, REQUEST2ACTION, EXCEED, ODYSSEE$_M$URES* projects.

[3]https://smartcities-infosystem.eu/sites-projects/projects/insmart.

tion (POE) surveys, monitoring of the implementation of specific laws and regulations such as registers of energy performance certificates and heating, cooling and air-conditioning systems inspections under the EPBD[4], and policy measures, such as energy efficiency and renewable incentives such as grants, tax deductions, energy efficiency titles, energy audits and meters under the EED, energy management systems, maintenance and operational cost monitoring, sensing and automation equipment in smart homes and smart buildings.

Data from these sources can be used to understand behaviour, assess performance, improve market competitiveness, monitor policy measures, planning and allocating resources and so on. As long as data availability is dramatically increasing, interpretation is imperative for:

■ setting data-informed goals

■ listing data sources and organizing them into useful pieces of information

■ gathering insights from data visualizations

■ tracking progress.

The national census of population and dwellings[5] in EU countries already provides an interesting basis for social and infrastructure analysis in buildings. EPBD repositories, used mainly for logging Energy Performance Certificates (EPC) and inspection data, have been managed for several years and continue to grow. Thus, these databases are increasingly being used not only for control and compliance goals, but also to complement other sources with the aim of monitoring the energy performance of buildings, enabling policies and supporting the decisions of market players [84].

BIM is a data-driven holistic modelling system that is based on 3D CAD. BIM data combined with building performance under occupancy can produce more accurate models. Instead of exchanging BIM data with the project parties only, as currently in use today, open BIM is now seeking to create and store data in a central repository with a single reference model. Building energy management systems (BEMS) generate large quantities of digital data. The open publishing of such data may share valuable company assets that can be leveraged by competitors. Data managed by utilities resides on multiple supports: bill data can be on paper, scanned images, PDF files or HTML pages on the utilities website, as text XML files. Moreover, interval meter data can be managed and provided to customers in different ways.

Nowadays the measurable, transparent, synergetic transition to low-carbon and more sustainable and smart cities, requires integration of building design and operation on a wider scale of urban energy districts combining technological and

---

[4]Directive 2010/31/EU on the energy performance of buildings (recast) - 19 May 2010.
[5]Directive 2012/27/EU on Energy efficiency (EED).

social aspects. Aggregated services for smart buildings and districts that integrate energy efficiency, economy, safety, mobility, flexibility of buildings in the electric grid, require a methodological definition of open ICT platforms (Smart District Platforms) that should enable multidisciplinary approaches making use of FAIR (findable, accessible, interoperable, replicable)[6] data.

Smartness implies increased empowerment of citizens to self-manage a series of functions connected to the energy network (towards energy communities) but privacy concerns, cybersecurity and social implications still unknown slow down the policy action.

8.3   Building data services: stakeholder engagement and value for users

Energy big data is meaningless unless its value is explored and mined, to support either the business decisions made by and customer services provided to private or public organizations.

For energy products (that is, HVAC, envelop components) and service providers, the value typically translates to developing more competitive marketing strategies. For customers the value comes in energy savings, efficiency, and improved visibility into how they are using energy.

Actual effectiveness requires users to be able to understand the new data structures, as well as the changes and opportunities that arise from big data. This will affect, for example, cost estimators and real estate managers, estate agents and brokers, urban and regional planners.

In Europe solution-oriented dialogue and integrated design with key stakeholders are hence required to make data available and useful, notably to boost the quality and quantity of building renovation towards local, national or supranational climate and air pollution reduction goals.

In 2017 the EU Intelligent Energy Europe Request2Action[7] project developed new and inedited services to make socio-economic and building energy data available to the supply and demand side market actors. To this extent a methodology was developed following the Product-service systems (PSS)[8] approach. The added value of this approach was co-creation with the potential users in order to attract and retain them as active contributors to the service performance.

---

[6]https://ec.europa.eu/info/sites/info/files/turning_fair_into_reality_0.pdf.

[7]https://ec.europa.eu/energy/intelligent/projects/en/projects/request2action.

[8]PSS are defined as a marketable set of products and services that are capable of jointly fulfilling customers' needs in an economical and sustainable manner [364].

Table 8.1: Distribution and number of stakeholders engaged.

TOTAL Number of stakeholders involved	226
Building owners	7
Energy Agencies and Advice	28
Policy makers and Local Authorities	75
Representative bodies of trade/professionals	47
Universities, Research Organisations	18
Investors and developers	9
Single traders and professionals	42

Various stakeholders were consulted and contributed to fix key requirements for data-services which provided added value of using energy and building data for planning and decision making and evidenced missing data for future service enhancement [85].

Table 8.1 quantifies stakeholders engaged in these data-service development in the REQUEST2ACTION project countries (Austria, Belgium, Greece, Italy, Netherlands, Portugal, Slovakia, United Kingdom).

This EU project found that:

■ Building-related energy databases integration is needed mainly from policy makers at different territorial levels (municipal, regional, national) aiming at implementing and monitoring building energy performance codes/policies and energy planning. Evidence-based information appears to be particularly interesting for cities to draft SEAPs (Sustainable Energy Action Plans) within the Covenant of Mayors initiative.

■ Distributors and providers are interested in spotting areas for potential deals.

■ Trades unions and larger traders or corporations would access data to expand and develop their business or targeted services, while small traders (SMEs) do not appear so interested and fear they might be affected by "unfair competition" from larger disreputed companies who have easier access to data.

■ Value for private companies resides in optimization of energy procurement contracts, sustainable reporting, market risk management, compliance with building benchmarking legislation (that is, on renewables deployment).

■ Banks do not trust calculated energy performance data (from EPCs) to issue loans and are not interested in tools uniquely based on that data.

■ Easy-to-interpret data was generally welcomed, notably geo-referenced maps and information systems (GIS).

Enterprises rank energy data collection & reporting as either "very Important" or "Important" but many of them find the process difficult and hard to execute. Lack of standardization (data definitions, billing cycles, tariff codes, and billing methods can vary from provider to provider) and isolated information (data in various organisations, departments, offices even in the same utility) are considered as the main barriers to this aim (Verdantix 2013).

8.4 Examples of integrated building data services

In 2016 the Italian national energy agency[9] mainly collaborated with the Lombardy in-house energy agency[10] developing a tool [98] based on integration of EPC data with the national Census and building renovation incentives data, thus establishing a relationship between estimated energy performance in EPCs, recent systems and materials installation, climatic and social data. Data in the DIPENDE tool are displayed in Fig. 8.1.

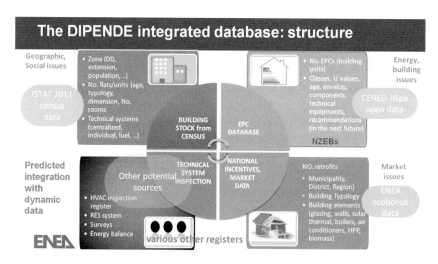

Figure 8.1: DIPENDE pilot data service - Structure of data.

The tool allows different users to estimate retrofit trends and select priority areas to support renovation strategies. The information was aggregated at municipal level and released in tables, graphs, maps and through a GIS. Records are

---

[9]National Italian Agency for New Technologies, Energy and Sustainable Economic Development, ENEA, www.enea.it.

[10]Lombardy region in house society Infrastrutture Lombarde S.p.A. (Ilspa), http://www.ilspa.it/home.

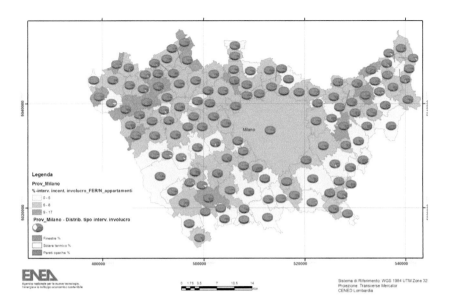

Figure 8.2: DIPENDE pilot data service - Retrofit market data display in Milan province.

aggregated at the urban level. The database covers more than 1,500 municipalities in Lombardy region, including around 70 fields for each record. Figure 8.2 shows the distribution of windows, solar thermal and the opaque building components retrofit compared to the total percentage of units undergoing any kind of retrofitting in Milan province (2012).

Windows replacement (blue sector in the cake) and deeper renovation (red sector) look almost independent of the building typology, quite different from the town (Milan) and hinterland. This information should guide targeted local incentives driving deeper renovation where the renovation market is still unfilled as well as pinpointed marketing by producers and installers.

A webGIS[11] was generated from the same integrated dataset, that allows multi-criteria analysis. The DIPENDE data service can be enriched with sectoral final energy consumption data, available on a monthly basis up to the province level. The consumption series will allow us to interpret how energy demand in the sector is evolving in quantity and mix, and evaluate whether targeted energy and climate saving policies are giving the expected results.

In the United Kingdom (Scotland), the Energy Saving Trust (EST) managing the EPC register for Scotland added value to the EPC data by cleansing it, addressing erroneous records and systematic biases in it and by statistically modelling an EPC value for those properties that do not currently have a real EPC.

---

[11] http://dipende-ba.casaccia.enea.it/pmapper/map_default.phtml.

This data is combined with 10 other datasets to create a comprehensive, reliable, up-to-date profile for 100% properties in Scotland. Geo-spatial modelling on property type, roof orientation, building size, garden size and other variables have been used as input to feed further analysis (for example, renewable energy resources potential). Through an arcGIS geographic information mapping and analytics software the local authorities are able to see the energy performance of their building stock at an address, community or whole-region level. As part of the REQUEST2ACTION project, EST developed a new service ("the Local Homes Portal"[12]) to provide the EPC data to the Scottish supply chain taking advantage of similar techniques to those already used with the local authorities. The data provided enabled companies selling energy services to identify which homes in which area will benefit from different energy saving measures.) The targets are householders, supply chain actors, local authorities, community groups, researchers and national policy makers. Local authorities are allowed by the government to access data to address at the level of individual buildings. In order to guarantee privacy and to avoid door-to door selling, data to consumers and to the supply chain is released at the local community area level (clusters of homes). This geographical boundary usually contains between 500 and 1,000 homes, identified by the postcode or council, and avoids privacy problems linked to eventual door-to-door selling.

For consumers, the web maps allow comparing the energy efficiency rating, energy consumption and carbon footprint of their homes to the average home in the same area. The service allows a view of aggregate and address level maps (limited rights service), view of filtered data, export to excel/csv, queries. (Costanzo et al. 2017). Aggregated and detailed view examples are illustrated in Fig. 8.3.

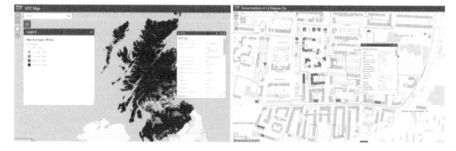

Figure 8.3: The EPC-based data tool in Scotland for home owners, local authorities and trades-people.

---

[12]https://localhomesportal.est.org.uk/.

More recently the EU ExcEED[13] platform, developed by five European organisations with ample expertise in data monitoring and analysis, supports building users and professionals through a European database collecting data from buildings, a front-end dashboard with integrated analytical tools, a set of tools that allows geo-clustered, statistical and knowledge analysis of building data, a benchmarking function to analyse building interaction, and an Indoor Environmental Quality (IEQ) survey to evaluate indoor comfort.

As users add their buildings' monitoring data, the platform transforms this data into knowledge using energy performance indicators and air quality surveys. The database is a combination of metadata and measured data. The measured data can be imported from a number of data sources including utility meters (usually provided by data collectors/aggregators), grid data (for example, electricity market data from market operators), monitoring data stored in csv files [113].

To help policy makers with planning and monitoring the platform would need a consistent amount of building cases. As the quantity of building data uploaded to the platform increases, benchmarking and aggregated results become more and more relevant. Starting from this experience an EU common and recognised buildings and districts data platform, harmonising and streamlining the data collection process, would allow the tracking of policies and performance to achieve the EU clean energy targets.

8.5   Towards more dynamic and automated collection of data on buildings

Metering generates millions of data on energy consumption. The comparison between calculated and real performance can add high value to the services that have been described so far.

As an example, the INSMART[14] EU funded project focused on data gathering and data processing tools and models, covering a wide range of city energy consumers, such as residential buildings, transport and utility services. A bottom-up approach was taken to gather data for the whole energy system from supply to distribution to demand, through a combination of different data sources. The project carried out the integration of multi-scale data granularity, including 15 minutes electricity consumption data from smart meters, monthly to annual statistical energy production and consumption data, detailed energy modelling of

---

[13]http://www.exceedproject.eu/register-to-the-platform/.

[14]FP7 INMART - Integrative Smart City Planning (2013-2016) bringing together four EU cities: Évora (Portugal), Cesena (Italy), Nottingham (United Kingdom) and Trikala (Greece), and scientific and technical organizations to establish a methodology for enhancing sustainable energy planning for future city needs through an integrative and multidisciplinary approach. https://smartcities-infosystem.eu/sites-projects/projects/insmart.

buildings typologies, city transport characteristics and mobility flows, and door-to-door surveys. The participation of different city stakeholders (municipality, utilities, transport companies, citizen groups, market associations) allowed to produce GIS tools to get insight from this data. In particular, in Évora, Portugal, the electricity consumption registries from smart meters have been linked to 72% of the sampled households to detail electricity consumption information. Consumptions according to number of persons per household, building typology, income levels, can be used to input the integrative planning tool for future energy policies evaluation: future energy scenarios have been produced per district and city area using TIMES Merkal models [140].

At a commercial level, Siemens currently integrates IBM's software, which includes Internet of Things (IoT) data analytics and asset management, into its cloud-based energy and sustainability management platform (the Advantage Navigator) in order to enable facility managers to benchmark buildings based on performance, forecast operational budgets, and predict potential equipment failures, among other functions. Weather monitoring and forecasting functionalities in IBM's cloud-based platform allows companies to better manage systems (for example HVAC) within one or several buildings around the world [232].

8.6   Conclusions

Big Data is a key enabler for reliable and effective policymaking, through statistical, business intelligence and predictive models. Collecting and making reliable data on buildings available can lead to enhanced consumer information, effective financing schemes to improve not only building performance but also quality of life in districts and cities.

Some European initiatives tried to manage large data related to current building energy performance. These actions incorporated the development of integrated databases and web oriented services, so that they are easily accessible and can attract more users to support energy retrofitting. In some cases concern on the impact of making data available to the private sector suggested limiting access only to those organisations contracted by central or local governments or to registered social landlords.

Integration of different existing data-sets may be complex and efforts remain rather uncoordinated in most countries. The main limitations are due to regulatory and juridical framework constraints.

The smart city vision holds out the promise of integrating data from multiple organisations, diverse environments, and a wide variety of intelligent devices. Yet, data integration even within organisations is one of the hardest challenges.

The adoption of open standards across the IT and communications industry can reduce (although not remove) the technical barriers.

To process static and dynamic data from multiple and diverse sources, by multiple and diverse processors (human or hardware) gathered through various means, a new approach is needed. This "collaborative" building data processing for decision support has not been mainstreamed and systematized yet [212]. A common vision towards innovative big data approaches is currently being used by the EU within the H2020 programme.

Integration of dataset formats and architectures and compatibility with smart meters, sensors, IoT devices, BEMS outputs should be facilitated by standardisation and simplified access. This might lead to new data-driven business models resulting in improved building performance.

Chapter 9

Selecting Suitable Plants for a Given Area using Data Analysis Approaches

Charitha Subhashi[a] and *Malka Halgamuge[b,*]*

9.1 Introduction

Australia has been considered as one of the top countries concerned with environmental issues that have not only promoted green living nonetheless has also become a leading country on sustainability[1]. While developing the country economically, implementing a sustainable and a Greener environment to its citizens should be the priority of every government. According to the statistics[2], Australia has approximately 297,857 registered plants. The recent scheme developed by the 'Natural Heritage Trust' shows that there are 30 Major Vegetation Groups in Australia that consist of 67 Subgroups[3]. Furthermore, to focus on the selected geographical area, community and council officers of Geelong, a local government area in the Barwon South West region of Victoria, has raised a few key

[a] FlexRule, Camberwell.

[b] The University of Melbourn.

* Corresponding author: malka.nisha@unimelb.edu.au

[1] https://www.usnews.com/news/best-countries/green-living-full-list.
[2] https://www.anbg.gov.au/aust-veg/australian-flora-statistics.html.
[3] https://en.wikipedia.org/wiki/Flora_of_Australia.

concerns[4] in it, such as establishing biodiversity protection, encouraging indigenous vegetation planting and environmentally sustainable landscaping through enhancements and incentive schemes. They have given a very special attention to biodiversity and plan to establish more plants in gardens. However, in order to minimize the risk of planting geographically unsuitable plants, it is necessary to establish and recognize suitable plants for each area for sustainability purposes.

Environmental compatibility of trees also needs to be accommodated when growing plants together in an area. Certain species can be grown in closer proximity with each other while some species show less growth rate compared to some group of species [301]. For instance, if a certain species has a root system that is interrupting each other, then it can have detrimental effects on the growth of both plants [301]. It is same with tree structures as well, in terms of height, shape, body, for example. It is evident that reconciliation of genes and species [300] is a major point that should be noted when selecting plants to be grown in a particular area. Through analysis, determining dominant and subordinate species of each area can provide a better guidance in creating a biodiversity in a more mindful manner.

One of the major ground level concerns in meteorological conditions is the urge to proliferate and bio diversify plants. Extreme weather conditions can put stress on the ecosystems which inadvertently has negative effects on native and exotic plants that leads to the loss of vegetation [249]. This also exerts pressure on declining species and raises questions about different areas and how they influence species' physiological tolerances [118], limitation of food resources [350], completion of life cycles and procreation of species and so on [377]. However, even in extreme weather conditions native plants show better performance in growth rates, and the plants are able to survive in extreme weather [249]. According to a previous research [118], "detect, and diagnose species that are declining at an earlier stage, is a challenging task that most conservation practitioners face". Although the research was carried out on different species, it talks about how plantation is associated with animal habitats and food chains. Therefore, it is important to detect which plant is more prominent in each area, to help identify and halt the degeneration of plants and other species.

As plant species related concerns directly effect all other species, there is a fundamental need to collect data on each plant species. Especially with the use of new tools and techniques, it becomes even easier to collect and store data. For instance, there are techniques to measure the height of trees by using a digital photo [394] and to measure tree crowns using satellite images [206]. Also, researchers have analyzed data and have identified patterns and interrelationships among tree measurements, as researchers physically take a lot of measurements and logically determine them. For instance, determining tree height using DBH

---

[4]https://geelongaustralia.com.au/.

(Diameter at Breast Height) can be taken (DBH–H ratio) [339]. As a result, the datasets get larger in a short period of time and the available data needs to be sorted by using data mining techniques. Data mining or Knowledge Discovery is the process that identifies patterns and builds relationships from an abundance of raw data through various analyses techniques in order to transform them into a usable meaningful format [294]. Moreover, it can also be classified into two major categories depending on the nature of the output, Predictive Data Mining and Descriptive Data Mining [144]. Predictive Data Mining calculates the future expected value by using a given set of data. Descriptive Data Mining builds the patterns and relationships with the available data. Additionally, depending on the way it presents the output, data mining allows data to be categorized into six groups; Anomaly Detection, Association rules learning, Clustering, Classification, summarization, and Regression [294]. Also, both Predictive and Descriptive Data Mining methods are used to get results for making decisions, and in order to do this clustering and classification techniques are also used. Clustering is an analytical method that groups closer objects in a distributed dataset [13]. Classification facilitates the prediction of the class of instances as it learns from a given set of pre-labelled data [239]. The chapter will discuss predictions that can be taken into account in order to determine the suitability of a tree plantation in a certain area. It will also give a clear idea about reducing the risks of growing insufficient and unsustainable plants in urban gardens areas.

The main contributions of this chapter include the following:

■ Carry out a study on plant species and their growth and health to select one to grow in a specific location for an agricultural purpose, gardening or for any industrial plantation and to identify declining plant species in each area.

■ Analyze the dataset consisting of 104,352 plant species planted in the city of Greater Geelong council region.

■ Determine the most suitable plant species for a given area using K-means algorithm.

■ Determine the plant health status using multivariate analysis.

■ Compare classification algorithms (that is, J48, JRIP, OneR, PART, Bayes Network, and Naive Bayes) to determine the plant health status and provide a classification using J48 as it showed the highest correctly classified percentage and lowest root mean squared error compared to the rest of the algorithms.

The chapter is organized as follows: Section I introduces the dataset including its attributes, and how the data is collected, background and motivation behind the research. Section 2 further discusses related works, their strengths and necessary improvements. Section 3 describes the used data collection methods, algorithms, and statistical analysis methods. Additionally, the results are presented in Section 4 with the analysis and tests carried out followed by the related discussion in Section 5. Finally, the chapter concludes with Section 6.

9.2 Related work

There are previous studies carried out focusing on various topics such as detecting declining species and improving species distribution. Therefore, in this section we will be discussing further information on related work.

Forest site productivity is an indicator used to describe how good a given forestland is and it depends on various variables such as soil, temperature, tree diameter and height. A research carried on how scaling in predictive modelling effects site productivity explains methods to be considered when determining the suitable plants for a specific area and the difficulty of scaling those methods. However, the attributes they are using are varied from our research and it is a process-based model which requires collecting data time-to-time over a certain period. The research is more focused on variables such as soil reports, plot coverage and age. Furthermore, the conclusion of this paper explains that process-based models can be less accurate and more complex.

Another study conducted on modelling approaches to estimate site productivity of uneven-aged forests by comparing height-age and diameter shows how height, age and DBA are connected to each other when predicting the forest site productivity. The research was carried out in northeastern China using natural Mongolian Oak (Quercus mongolica Fisch.) and Korean Larch (Larix olgensis Henry.). Its results illustrate that the prediction accuracy of dominant height–dominant diameter is higher than dominant height–age model. However, the study does not consider the facts such as structure and health status of trees. Also, when considering the site productivity predictions, it is not capable of determining suitable plants to get the maximum productivity as it only considers the overall woodland productivity.

Another research done on "genome mapping and marker-assisted selection in aquaculture" explains on how to select suitable species for aquaculture, which is a thriving area in agriculture. The method they discuss has used over 40 species and have mapped quantitative trait loci (QTL) for over 20 species traits to create a linkage among them. The method has been applied for aquaculture and there is no evidence that the same linkage can be used in land agriculture to select the suitable plant species. The genome-wide association study which was used in the paper can be more accurate although it requires more advanced and expensive techniques showing the need for alternative data mining methods.

The importance of data mining is huge, and there is previous research that exudes on plants classifications. A research carried out in Finland has collected species data from "an urban garden in the City of Espoo, Finland" showing the importance of plant identification [287]. They have scanned the plants through mobile laser scanning while collecting hyperspectral data to classify trees. It requires the knowledge of using the Sensei system, that they used in their method which can be a limitation when considering a larger area to select plants to be grown. Finally, combining the two, they have been able to correctly classify the plants. The study was more focused on urban, park and road-side plant planning and management, and agriculture sector is a potential area of interest.

Another study carried out in Flanders, Northern Belgium [10] has mapped species in a forest area by using different techniques: "(a) location-based technique (ordinary kriging), and (b) attribute-based technique (regression) and three hybrid techniques (Geo-matching, Ordinary co-kriging, and Regression Kriging)". At the end of the research they discuss different mechanisms that can be used to determine the forestry site index. Their classification of the species in the area, considers attributes such as number of specimens of each species, dominant height of a plant in a selected plot and age to create the forestry site index. Our approach uses location as well as attributes. However, in our research, more attributes such as health status, and DBH are considered. Also, there are online tools known as, "Plant Selector+"[5] which have been created to be used for South Australian plants. Once the suburb name or postcode is entered it shows a list of suitable plants. The method they have used to identify the plants are not mentioned. Nonetheless, it is specific to South Australia and considers the suburb or postcode instead of a location which is more accurate. Therefore, it cannot be applied to another location or area to be used.

According to the review conducted on previous work, it is necessary to research a large amount of plant data in a given area and finding a method to select plants that can also be applied to any other geographical area in the world. It will provide a guideline on how to select the plants and what criteria can predict the suitability of plants.

9.3   Materials and methods

This paper has used a predictive method to predict the expected outcome. Also a few clustering algorithms: K-means and classification methods: Bayes Network, J48, JRIP, Naive Bayes, OneR, and PART are used.

---

[5]http://plantselector.botanicgardens.sa.gov.au/home.aspx

9.3.1 Data collection

The dataset was taken from Data.gov.au, the official website of published datasets of the Australian government. The dataset has been published in 2015 by the City of Greater Geelong Council, Victoria, Australia. It contains details of trees in Geelong such as species, height, and health (Good, Fair, Poor, Dead).

9.3.2 Data processing

Figure 9.1 shows a step by step method used to transform raw data into a meaningful result, in order to understand suitable species that can be grown in each subarea.

The original dataset contained 21 attributes and 119,168 instances and was not completely readable. It was processed before the analysis. For instance, there were rows containing only the species name without any other data and they were removed from the dataset. Then 104,352 instances were selected for the analysis and they were grouped into 5 clusters according to the geographical area of the species. Next, the attributes were selected and the dataset was analyzed. In the end, after performing an evaluation, the results of the analysis were finalised.

9.3.3 Data inclusion criteria

Certain attributes such as collection date, location, address and reference numbers were removed as they were not related to the focused study. Furthermore, the instances with missing values were also removed. Therefore, at the end, 9 attributes mentioned in Table 9.1 were selected for research purposes and after pre-processing, 104,352 instances were used.

9.3.4 Algorithms

Following the steps of the flowchart in Fig. 9.1, three different algorithms were written for three major flows.

Algorithm 1 shows how the results were gained manually using multivariate analysis, Algorithm 2 shows the steps of K-means clustering and Algorithm 3 analyses the dataset using classification.

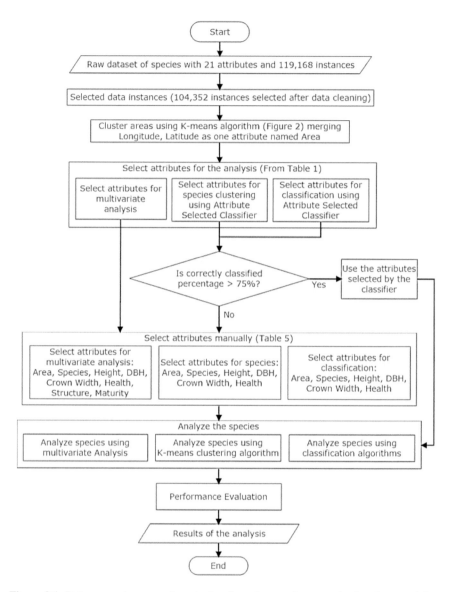

Figure 9.1: Data processing procedure starting from the raw dataset, selecting data, used data mining algorithms for the analysis of the evaluated results.

The multivariate analysis was carried out in Algorithm 1 compares each attribute against the number of plants in each area. For instance, in each area which plant species has the highest number of plants that comes under the category of "good" health status. This result will also be helpful to evaluate the results generated from data analysis methods.

Table 9.1: Descriptions on the selected nine attributes of the analysis.

Attribute Name	Data Type	Description
Longitude	Numeric	Longitude value of the location and coordinates of a specific tree
Latitude	Numeric	Latitude value of location coordinates of a specific tree
Species	Nominal	The class of the tree
Height	Numeric	height of the tree from top to bottom (in meters)
DBH (Diameter at Breast Height)	Numeric	A standard method of measuring the diameter of the trunk of a standing tree (in meters)
Crown Width	Numeric	Size of the mass of foliage and branches growing outward from the tree (in meters)
Health	Ordinal	Pre-determined health status of the trees. Options: Good, Fair, Poor, Dead
Structure	Ordinal	The hierarchical nature of the tree from top to bottom Options: Good, Fair, Poor, Failed
Maturity	Nominal	Different maturity levels of the plant Options: Mature, Semi-mature, Young, Developer, On Maintenance

Algorithm 1: Multivariate Analysis

string [] attri_m → Attributes for the analysis
string s_m → The most suitable species for the area

get a dataset of plant species in a specific area
compute area clusters using K-means algorithm
determine attri_m for multivariate analysis manually
compute the s_m performing multivariate Analysis using attri_m
print s_m

Algorithm 2 provides the steps to analyze the dataset using K-means clustering. The dataset was divided into a multiple number of clusters and during each cycle, the centroid plant species was selected. This was later analysed to see which plant is dominant in each cluster and cycle. For instance, when k=3 (k is the number of clusters), in each cluster which plant is the dominant plant.

Algorithm 2: Analysis using K-means clustering

string [] attri_clu → Attributes for the analysis
string [] species_selected_by_clustering → Species selected by clustering
string s_clu → The most suitable species for the area
int c → Correctly classified percentage
int a% → Pre-determined correctly classified percentage
int k → Counter for number of attributes
int num_species_selected_by_clustering → Counter for number of species
selected by clustering

get a dataset of plant species in a specific area
compute area clusters using K-means algorithm
determine attributes for species clustering using Attribute Selected Classifier
if c >astring [number of attributes] attri_clu = attributes selected by the classifier
else attri_clu = attributes selected manually
end if
while (k= number of attributes) **do**
if k=1 **then**
while (k= number of attributes) **do**
num_species_selected_by_clustering = num_species_selected_by_clustering+1
end while
define string [num_species_selected_by_clustering] species_selected_by_clustering
end if
string [num_species_selected_by_clustering] species_selected_by_clustering =
species selected by K-means
clustering algorithm using attri_clu
end while
k=0
for (k= number of attributes)
get the count of each species in species_selected_by_clustering
end for
s_clu = maximum count of single species
Print s_clu

Algorithm 3 shows how the dataset can be analyzed using classification algorithms. Six classification algorithms (Bayes Network, J48, JRIP, Naive Bayes, OneR, and PART) were used for testing and for each algorithm, three methods were used; full training set, cross-validation with 10 folds and percentage split. For each test, evaluation criteria such as correctly classified percentage and root mean squared error were determined to identify the best classification method.

Algorithm 3: *Analysis using classification algorithm*

string [] attri_clu → Attributes for the analysis
string [] classification_results → Recommended measurements
(height, crown width)
string [] classification_methods → Selected classification methods
int pcc% → Correctly classified percentage of classification method
int highest_pcc → Highest correctly classified percentage
int c → Correctly classified percentage
int a% → Pre-determined correctly classified percentage
int k → Counter for number of attributes

get a dataset of plant species in a specific area
compute area clusters using K-means algorithm
determine attributes for species clustering using Attribute Selected Classifier
if c >a% **then**
string [number of attributes] attri_cla = attributes selected by the classifier
else attri_cla = attributes selected manually
end if
while (k = number of classification methods) do
[] classification_results = results of the classification method performed from
[] classification_methods
[] pcc% = correctly classified percentage
end while
highest_pcc = highest percentage from [] pcc%
s_cla = the result from [] classification_results related to highest_pcc
Print s_cla

Pre-processing is highly essential because pattern recognition depends on the data that is being fed to the system. First, the attributes were selected from raw datasets to get the cleaned version. Next, the fluctuation of some numeric values was high and certain values were available as a range (for example, >20m). Therefore, the values were grouped as 0-5, 6-11 ... n- (i+5) where n=j*6 and j=0, 1, 2 ... and the average value was taken from each group. The transformed dataset was sent as the input for each data mining algorithm.

Table 9.2 shows the selected classification and clustering of algorithms along with their description.

Table 9.2: Clustering and clustering algorithms of the analysis used to generate the results.

Algorithm Name	Algorithm Type	Description
Simple K-Means	Clustering	This function groups the dataset into predefined k number of clusters by initializing centers of each cluster [388]. It was used only to cluster the dataset according to geographical locations.
Bayes Network	Classification	Bayes Network is a statistical model that uses probabilistic graphical model. In this directed acyclic graph (DAG), nodes of the network represent random variables and arc represent the connections among variables [388].
J48	Classification	This function uses C4.5 algorithm to build a decision tree and also uses the ID3 principle to build decision trees [272].
JRIP	Classification	JRIP generates "a rule learner and incremental pruning to produce error reduction (RIPPER)". It is an optimized algorithm to classify instances [68].
Naive Bayes	Classification	Naive Bayes is based on Bayes Theorem and it assumes the independence among predictors in order to classify probabilistically [223].
OneR	Classification	OneR generates a rule for each parameter against the predictor and selects the rule with the least errors [190].
PART	Classification	PART is a combination of divide and conquer strategy. In each step, a C45 tree is built incompletely and the best leaf the rule is used to complete it [107].

9.3.5 Data and statistical analysis

The main concern of the research is to apply the clustering and classification methods to generate outputs. The various methods used for multivariate analysis show that species clustering and classification will not only be helpful to determine the suitability of plants for different areas in the city of Greater Geelong council, nonetheless it also helps to select which algorithm is better depending on correctly classified instances and time that was taken.

Open source software is used for the application of algorithms which is Weka (Waikato Environment for Knowledge Analysis, version 3.8). This tool has an in-built machine that uses algorithms with visualization tools and easy-to-use interfaces. The computer processor is Intel Core i5-3337U, 1.8GHz, and RAM (Random Access Memory) was 4GB.

It is important to perform a statistical hypothesis testing by calculating the P-value to statistically prove that the selected species are truly suitable to each area and this value should be less than 0.05 (P-value >0.05). P-value is the probability of gaining an outcome similar to or so extreme than what was actually observed when the null hypothesis is true. P-value among the areas was 0.014 and it was calculated using One-way ANOVA test in IBM® SPSS® Statistics 20 which was designed for hypothesis testing.

The null hypothesis of this research is that there can be one species for a particular area more suitable than the other species. The alternative hypothesis is that all the species can be grown in all the area.

H_0: There is one species for a particular area more suitable than the other species
H_1: All the species can be grown in all the areas equally

Basically, at the end of the test, it should prove that the value of Mean Square error within groups is lesser than the value of Mean Square error between groups.

9.4 Results

In order to obtain results on plant suitability, predictions and performances, different model evaluation techniques such as full training set, cross-validation and percentage split were used in this study. The main steps of this section were illustrated in Fig. 9.1 from the step, 'Cluster areas using K-means algorithm merging longitude, latitude as one attribute named area' to 'Analyze the species'.

The areas were clustered according to the geographical location coordinates (longitudes and latitudes). Table 9.3 shows a few clustering algorithms used for clustering.

Table 9.3: Tested clustering algorithms to group the geographical data into clusters.

Clustering Algorithm	CPU time	Number of Iterations	Within cluster sum of squared errors	Log likelihood
Canopy	0.81	-	-	-
K-means	4.8	46	1957.37	-
Farthest First	0.36	-	-	-
EM	79.06	47	-	3.37
*Cobweb	2212.65	-	-	-

* Not able to define the number of clusters

9.4.1 Area clustering

K-means clustering shows a less CPU time although it is the not the least time. Also, it shows the squared errors, which is one of the important performance evaluation techniques. Therefore, at this stage, K-Means clustering algorithm was used to cluster the coordinates into five clusters as Fig. 9.2.

The idea of clustering the area is to identify the most planted trees in each area. Instead of studying separate locations, it is practicable when looked into areas. After clustering the areas, attributes were selected to start the analysis of plant species.

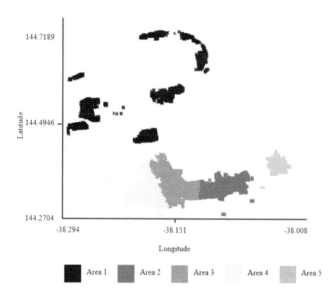

Figure 9.2: The five area clusters of the plant distribution using K-Means clustering algorithm.

9.4.2 Selecting attributes

The very first step, was to select necessary attributes of the dataset as there were unrelated ones such as reference numbers and data collection date, among others. In order to select attributes of the dataset, the ones selected were tested along with different classification algorithms. Table 9.4 shows the attributes selected by this classifier.

According to Table 9.4, the percentages of correctly classified instances are very low. Most importantly, in order to make proper decisions on tree heights, DBH,

Table 9.4: Attributes selected by the attribute selected classifier applied with each classification algorithm.

Classification Method	Selected Attributes by the Attribute Selected Classifier	Correctly Classified instances	Root Mean Squared Error
Bayes Network	height, DBH, structure	62.84%	0.3188
J48	height, DBH, structure	63.73%	0.3229
JRIP	height, DBH, structure	63.69%	0.3237
Naive Bayes	height, DBH, structure	56.83%	0.3391
OneR	height, DBH, structure	63.37%	0.3828
PART	height, DBH, structure	63.74%	0.3166

* Not able to define the number of clusters

Table 9.5: Selected attributes for multivariate analysis, species clustering, and classification after area clustering.

Analysis method	Used Attributes
Multivariate Analysis	Area (derived from Longitude and Latitude), Species, Health Status, Maturity, Structure
Species Clustering	Area (derived from Longitude and Latitude), Species, Height, Crown Width, DBH, Health Status
Classification	Area (derived from Longitude and Latitude), Species, Height, Crown Width, DBH, Health Status

Crown Width, Health, Structure, and Maturity are important [40, 10, 375]. Also since the main idea of this study is to identify the species for each area, their names were also essential. Therefore, the attribute selection was done manually as the attributes selected by the Attribute Selected Classifier is not sufficient for this study. Table 9.5 shows the selected attributes. Finally, longitude, latitude, species, height, DBH, crown width and health were selected for the clustering, classification, and multivariate analysis was carried out using maturity and structure as well.

Finally, the analysis and the main focus of this study was carried out as follows.

9.4.3 Multivariate analysis of species

This section will discuss various analyses carried out with the selected attributes of the dataset.

Table 9.6 shows the number of species in each selected area. The huge variety of species makes it harder to decide which plant is more suitable to plant in each area. Therefore, further analysis was carried out as follows based on different criteria: species population, maturity, and structure.

Table 9.7 shows the planted species of each area along with the percentages. It is clearly visible that the percentages of commonly planted species are low. Therefore further analysis was performed based on the top three species in order

Table 9.6: Number of species and number of plants in each area cluster.

Area	Number of Species	Number of Plants
1	174	8298
2	196	15090
3	209	7646
4	242	33759
5	242	39559

Table 9.7: Number of species and number of plants in each area cluster.

Area	Mostly Planted Species	Scientific Name	Number of plants	Percentage
1	Ficifolia	Corymbia ficifolia	1121	13.51%
2	Cladocalyx	Eucalyptus cladocalyx	1226	8.12%
3	Canariensis	Phoenix canariensis	369	4.83%
4	Conifer	Pinophyta	2778	8.23%
5	Citrinus	Callistemon Citrinus	2899	7.33%

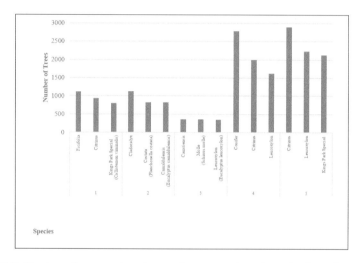

Figure 9.3: Number of trees against the top three most planted species in each area cluster (area 1, area 2, . . . , area 5).

to provide a more accurate decision. Figure 9.3 shows the top three most planted species in each area proving that Citrinus and Leucoxylon are very commonly seen plants in Geelong.

Figure 9.4 shows the percentage of unique species in each area according to their health status. According to that, each species has more than 90% of trees with good health condition. Only less than 5% of plants of each species are poor.

Figure 9.5 shows the number of top mature and semi-mature trees in each area. According to that, Citrinus, Cladocalyx, Canariensis and Confertus, are species with a large number of mature and semi-matured trees in each area.

Figure 9.6 shows top three species in each area with a good structure as Fig. 9.7 shows species with good structure within the whole of Geelong area. According to these facts each area, Citrinus, and Kings Park Special show a significant number of trees.

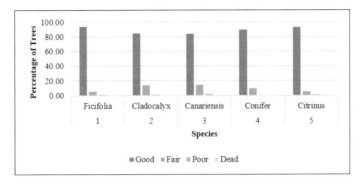

Figure 9.4: Percentage of trees against the most planted species of each area cluster (area 1, area 2, ..., area 5) grouped by their health status.

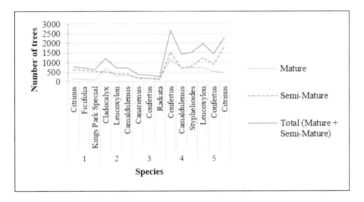

Figure 9.5: Number of trees against the most planted species in each area cluster (area 1, area 2, ..., area 5) showing their maturity levels.

9.4.4   Species clustering

With the expectation of clustering species, K-means clustering algorithm was used on species dataset. A number of clusters (k) were assigned from 1 to 6 as there are 6 attributes in the dataset. Table 9.8 shows the results of the tests.

According to Table 9.8, Citrinus is shown in all the clusters. Also when k≥4, this species repeats. Simultaneously, the raw data analysis in Section 3.3 shows that Citrinus is at the top of most of the analysis.

9.4.5   Classification

After carrying out the Multivariate Analysis of the species, six classification algorithms (Bayes Network, J48, JRIP, Naive Bayes, OneR, and PART) were used to make predictions. To test each algorithm, mainly three different testing tech-

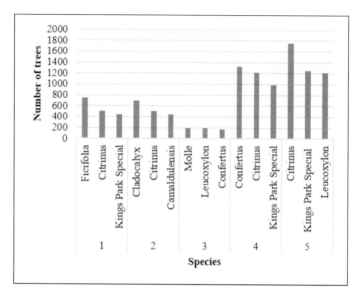

Figure 9.6: Number of threes with a good structure against the top three species in each area cluster (area 1, area 2, ..., area 5).

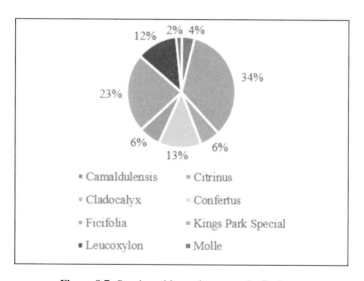

Figure 9.7: Species with good structure in Geelong.

niques were used: using full training set, cross-validation of 10 folds and percentage split secured the results.

Table 9.8: Species clustering using k-means clustering algorithm when k = 1, 2, 3, 4, 5, 6.

Number of Clusters (k)	Species Selected by K-means Clustering
k=1	Citrinus
k=2	Confertus, Citrinus
k=3	Confertus, Citrinus, Leucoxylon
k=4	Confertus, *Citrinus, Leucoxylon
k=5	Confertus, *Citrinus, Leucoxylon
k=6	Confertus, *Citrinus, Leucoxylon, Kings Park Special

* Repeated values observed in more than one cluster

9.4.5.1   Using full training set

The first attempt was to test the full dataset. This way, the full dataset is treated as the model and then it is used for evaluation. This can be misleading as good algorithms can store all training patterns and achieve perfect classification scores. Table 9.9 shows the results of this technique for each classification method.

Furthermore, the correctly classified instances can be divided as TP (True Positive) and FP (False Positive). Also, the incorrectly classified instances can be grouped into FN (True Negative) and FN (False Negative). To avoid the accuracy paradox, performance evaluation measurements were also used to measure the results and validate them.

The percentage of True Positive predictions: Precision (P)
The true positive rate (sensitivity): Recall (R)
The balance between the Precision and Recall: F-Measure (F)
Precision (P) = TP / (TP+FP)
Recall (R) = TP / (TP+FN)
F-Measure (F) =2*R*P/(R+P)

The results in Table 9 show that correctly classified percentages of J48, JRIP, OneR, and PART are quite significant at a confidence level of 91%. However, considering the CPU time, and the time it takes to build the model JRIP took less time than expected while J48 records took the longest time.

9.4.5.2   Cross-validation with 10 folds

K-fold validation with k = 10 was used as the next technique. The number of folds determines the number of partitions of the dataset structure. The first 9 folds (k-1) are used to train the dataset and the last fold is for testing. Finally, it calculates the average performances of all k models. This technique is the "golden standard" as it assesses the performance of a model. Table 9.10 shows the comparisons of the results gained in each classification method.

Table 9.9: Correctly and incorrectly classified instances, execution time and root mean squared error using full training set.

Classification Method	PCC (%)	MAE	TP Rate	FP Rate	Precision	Recall	F-Measure	CPU Time (seconds)	Root Mean Squared Error
Bayes Network	89.64	0.08	0.90	0.85	0.85	0.90	0.87	1.03	0.21
J48	91.10	0.08	0.91	0.90	0.89	0.91	0.87	11.05	0.20
JRIP	91.09	0.08	0.91	0.90	0.87	0.91	0.87	0.23	0.20
Naive Bayes	89.37	0.09	0.90	0.84	0.85	0.90	0.87	0.21	0.21
OneR	91.07	0.04	0.91	0.91	0.89	0.91	0.87	2.63	0.21
PART	91.16	0.08	0.91	0.90	0.89	0.91	0.87	2.94	0.20

Table 9.10: Measurement of detailed accuracy of each classification method using cross validation with 10 folds.

Classification Method	PCC (%)	MAE	TP Rate	FP Rate	Precision	Recall	F-Measure	CPU Time (seconds)	Root Mean Squared Error
Bayes Network	89.60	0.08	0.90	0.85	0.85	0.90	0.87	1.33	0.21
J48	91.09	0.08	0.91	0.90	0.89	0.91	0.87	12.01	0.20
JRIP	91.08	0.08	0.91	0.90	0.88	0.91	0.87	64.22	0.20
Naive Bayes	89.32	0.09	0.89	0.84	0.85	0.89	0.87	0.51	0.22
OneR	91.03	0.04	0.91	0.91	0.84	0.91	0.87	0.29	0.21
PART	90.99	0.08	0.91	0.90	0.86	0.91	0.87	260.12	0.20

According to the results in Table 9.10, J48 is the highest percentage of correctly classified instances which are 91.09% and Native Bayes has the lowest value, 89.32%. Considering the time taken to build the model, the PART algorithm has taken the highest execution time duration of 260.12 seconds to build the model while OneR algorithm has taken only 0.29 seconds. Also, it indicates that the prediction rates of each classification algorithm, J48 and JRIP show the best TP rate which is 91.1% and Naive Bayes shows the least which is 89.3%.

9.4.5.3 Percentage split

The dataset was tested by splitting them into different percentages. In this method, the model is trained to build-up with a certain percentage of data and then it is tested with remaining percentages with the rest of the percentage.

TITLE

I

Table 9.11 shows the correctly classified percentage of each classification algorithm. As a graphical presentation, Fig. 11.8 clearly illustrates the correctly classified percentage of each algorithm with different percentages of tested data. According to that, J48 shows a high percentage of accuracy while Native Bayes shows the least correctly classified rate.

Table 9.12 shows the ranges of measurements which show the highest probability for each Good, Fair, Poor and Dead categories. For instance, the species with heights up to 5.25 meters have a high probability of being good considering the health status. However, the table created using Bayes Network classifier does not show a clear difference between groups.

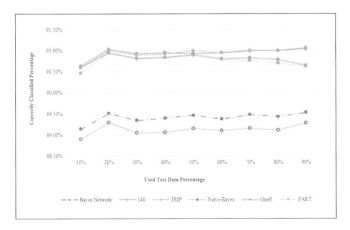

Figure 9.8: Comparison of correctly classified percentage of each classification method with percentage split technique.

Table 9.11: Correctly classified percentage and root mean squared error of each classification method with percentage split technique.

Classification Algorithm	Train 90%: Test 10%	Train 80%: Test 20%	Train 70%: Test 30%	Train 60%: Test 40%	Train 50%: Test 50%	Train 40%: Test 60%	Train 30%: Test 70%	Train 20%: Test 80%	Train 10%: Test 90%
Bayes Network	89.14% (0.2147)	89.51% (0.2114)	89.34% (0.2131)	89.40% (0.2133)	89.46% (0.2124)	89.37% (0.2132)	89.48% (0.2127)	89.43% (0.212)	89.53% (0.2115)
J48	90.64% (0.2063)	91.03% (0.2025)	90.93% (0.2036)	90.95% (0.2034)	90.93% (0.2037)	90.95% (0.2033)	91.00% (0.2028)	91.00% (0.2038)	91.04% (0.203)
JRIP	90.60% (0.2074)	91.00% (0.2035)	90.90% (0.2044)	90.92% (0.2042)	91.00% (0.2035)	90.94% (0.2041)	90.99% (0.2036)	91.0% (0.2035)	91.07% (0.2027)
Naïve Bayes	88.90% (0.2217)	89.29% (0.2183)	89.05% (0.2197)	89.06% (0.2207)	89.15% (0.2195)	89.10% (0.2206)	89.15% (0.2195)	89.11% (0.2199)	89.28% (0.2186)
OneR	90.60% (0.2168)	90.94% (0.2128)	90.82% (0.2142)	90.84% (0.2140)	90.90% (0.2133)	90.81% (0.2143)	90.83% (0.2142)	90.78% (0.2148)	90.65% (0.2162)
PART	90.47% (0.2079)	90.95% (0.2034)	90.81% (0.2047)	90.84% (0.2045)	90.89% (0.2041)	90.79% (0.205)	90.76% (0.2052)	90.70% (0.2067)	90.62% (0.2080)

Table 9.12: Probability distribution generated by Bayes Network classifier.

	Health Status			
	Good	*Fair*	*Poor*	*Dead*
Height (meters)	-inf-5.25	-inf-5.25	-inf-5.25	-inf-5.25
Crown Width (meters)	-inf-5.25	-inf-5.25	-inf-5.25	-inf-5.25
DBH (meters)	-inf-30	30-50	-inf-30	-inf-30

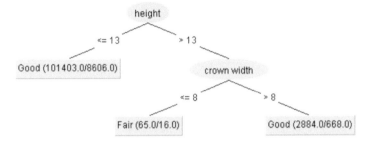

Figure 9.9: Visualization of probability of a good plant from J48 Classifier.

Figure 9.9 was created using J48 Classifier and it shows the possibility of a good and healthy tree. For instance, if the tree is taller than 13 meters and the crown width is wider than 8 meters there is a high probability of it having a good health status. The aim of this research is to identify the most suitable plants for the Geelong region. K-means clustering algorithm was used and different analysis methods were deployed to generate the best output while confirming the results in K-means clustering algorithm. Clustering algorithm shows that Confertus, Citrinus, and Leucoxylon are the best species to be planted (p = 0.014). Considering the number of plants in each sub-area under raw data analysis, Ficifolia, Cladocalyx, Canariensis, Conifer and Citrinus species are widely grown. The health of each species has also shown as more than 90% and plants are in good condition. Only a very few plants are in poor or in dead condition. As the diversity of fluctuation is high, the top three plants were considered for further analysis. The mature plants such as, Citrinus, Cladocalyx, Canariensis, Confertus and Leucoxylon are species that show a good amount of maturity as semi-mature plants.

Considering the classification of algorithms, J48 gives the best results with a clear connection among attributes. However, overall all six algorithms show good results. For instance, Fig. 9.8 shows that the fluctuation among the correctly classified percentages of algorithms is less than 2%.

9.5 Discussion

Plants "are the foundation of all terrestrial ecosystems" [244]. Alternatively, technology is updating and upgrading every day. Therefore, the importance of combining the importance of conserving plants, and technology, promises sustainability by identifying plants which are more suitable in certain areas than others. The paper discussed a method of optimizing the way of selecting plants. The main purpose of this study is to identify the suitability of plants for a particular area and an analysis was carried out using a dataset of 104,352 existing plants in Geelong, Australia. Establishing the foundations of this research, as an article in Arena magazine explains how the clash of ecosystems takes place due to industrial advancement. Additionally, they introduced plants and explained why it is important to relocate native species for sustainability [76]. In the selected dataset in some of the divided plots, there are more than 200 different species of plants that belong to all these categories. According to the analysis, there are 5 subareas that have 5 different dominant plants Ficifolia: 13.51%, Cladocalyx: 8.12%, Canariensis: 4.83%, Conifer: 8.23%, Citrinus (Callistemon Citrinus): 7.33%. However, these plants have an astounding ability to adapt to the environment, regardless of the state of being industrial, introduced or native, as the growth success, of a plant in a particular area can vary. The result of this research has also revealed that this variation is an important aspect that should be considered. For example, in subarea 3, Canariensis (Phoenix canariensis) has the highest number of matured and semi-matured trees although Molle (Schinus molle) is the best structure. Additionally, rather than merely assuming that native plants can grow better, the necessity of a research arises that can justify the suitability of species that grow in a specific location by means of existing plants' and biophysical indicators such as tree performances based on height, DBH, health, and more. It can be taken as an approach of adopting what has been understood when the circumstances change [76]. While the diversity changes and the necessity of its understanding arises, at the same time technology, assists people to find more accurate answers faster than ever. Therefore, using technology, machine learning algorithms give a better understanding of diversity. Machine learning can be defined as a data-driven method of constructing models by using indications acquired from a sample dataset and the output of a prediction, as a result of identified patterns of training phases [335]. There are three categories of machine learning styles: supervised (using labeled data), Semi-Supervised (using both labeled and unlabeled) and unsupervised (using unlabeled data) learning methods [117]. Machine learning has been used by many scientific studies for predictions in different fields [226, 15, 96, 291, 391] including environmental studies. The ability of feature extraction and anomaly recognition has become a major reason for using machine learning algorithms [291]. In this study, once the raw data is fed, using the K-means clustering algorithms it showed that Citrinus (Callistemon citrinus) is a species that is mostly planted, and raw data analysis revealed the same results evidencing precision (p = 0.014). Another advantage is that it is not necessary to have a domain knowledge when building a physical model in order to get the results from machine learning algorithms [96]. Therefore, even without a sound knowledge of plants, it is possible to use machine learning algorithm. However, its limitation is that it requires large amounts of data to give adequate results [263] and the quality of the predictions depends on the dataset. The variation of results given by different algorithms for the same datasets has become an issue in a previous study [291]. However, the results obtained by the current research shows only 2% of fluctuation among correctly classified percentage, proving that the results are insignificant in fluctuation. Also, the three classification testing techniques carried out proved that J48 gave results with the highest correctly classified rate (approximately 91%) and Naive Bayes resulted in the lowest (approximately 89%) although Naive Bayes showed the least CPU time to build the model. A research carried out to identify users' web movements and predicted requests have shown that J48 and Bayes Net have the highest percentage of correctly classified instances [202], although this study showed OneR and JRIP were better than Bayes Net. Another research mentions that CART is better than J48 [255] and this study did not test data using CART algorithm. Considering the CPU time, this study showed that J48 took more time, however, a different study mentions that J48 is comparatively faster [349]. Likewise, other studies have also shown that J48 algorithm is capable of providing a high percentage of correctly classified instances [190, 108, 323] although the time it takes to build the model is slightly higher than algorithms such as OneR, Naive Bayes. Furthermore, a study on decision tree algorithms [190] proves that decision trees such as J48 are easier and more powerful algorithms. Therefore, even without a vast knowledge of machine learning algorithms, it is possible to apply these algorithms in any study area. With the use of machine learning algorithms mentioned above, and plant species clustered and classified, the accuracy of clustering of raw data analysis was also being carried out. Moreover, while numerous studies have shown the importance of biodiversity [76, 359, 81] and smart farming [238], one of the large-scale researches on biodiversity reforestation emphasized the importance of mapping locations with the corresponding species and the use of biophysical indicators [81], yet it did not suggest a proper methodology to map species. Also, a study carried out in Brazil has done a slightly similar study to understand the dominant species dividing a selected forest into plots [276] considering only height and DBH by using K-means clustering [149, 373, 150], the area that was divided into 5 subareas. Another two studies carried out in Finland and in Flanders, Northern Belgium have done a similar research on mapping species. However, both studies only count the number of specimens of each species. Therefore, for the improvement of optimizing plant restoration practices, we moved forward in this research with the idea of using data analysis [180] with machine learning algorithms with a dataset that consists of tree measurements such as height, DBH, structure, and health status. For instance, K-means clustering algorithm shows that Citrinus, a native plant is very common in Geelong region, Australia (p=0.014). The raw data analysis also shows more than 90% of Citrinus plants are in good health and considering the structure and maturity of these species they are still at the very top.

9.6   Conclusion

Starting from a raw dataset of tree details, this paper investigated and found that, suitable tree species should be planted in a particular area to make predictions with the results of tested machine learning algorithms (clustering and classification) for sustainability. Also, various multivariate graphs were further analyzed to obtain more accurate and concrete views of the potential knowledge by comparing the accuracy of machine learning algorithm results. The results indicated that K-means clustering can be used to identify the dominant plants in each area and J48 gives the best results on prediction. However, all the classification algorithms tested have delivered prominently similar results showing less than 2% oscillation among correctly classified percentages. Also, as a matter of fact, the results depend on the dataset fed to the machine learning algorithms which can be taken as a limitation. Furthermore, the same research methodology can be applied to different geographical areas in order to find the most suitable plant species for the area. As a final point, the gained knowledge is important for the environment and plant related studies. The importance of knowing suitable plants for a specific area is helpful to understand the compatibility of plant species. It will lead planters, breeders, farmers and the agricultural industry to take better decisions before selecting the plants for a given area. A wider perspective of forestry, botany, and ecology will facilitate scientists to study more on the areas such as the tolerance of plants in different areas' certain weather conditions, selecting areas for studies on a specific plant, declining plant identification and dominant plant variation. Finally, it is the post-genomics era where the maturation of technology guides humans to understand plant species in a broader way and this study is a part of it.

Chapter 10

Ontology-Based Security Requirements Framework for Current and Future Vehicles

Abdelkader Magdy Shaaban,[a,] * Christoph Schmittner,[a] Thomas Gruber,[a] A Baith Mohamed,[b] Gerald Quirchmayr[b] and Erich Schikuta[b]

10.1 Introduction

Technology becomes an integral part of our lives; it changes how we communicate, travel, study, work, treat, entertain, and conduct other daily activities. Technology has created spectacular tools and machines to pave the way into a faster, comfortable, reliable, and more enjoyable life. The Internet plays an essential role in this progress. For example, the Internet of Things is considered the most advanced Internet technology [12]. The Internet of Things (IoT) is a huge network of interconnected units such as appliances, cell phones, sensors, cars, and people. This situation defines different relations between people-people, people-things, and things-things [260]. From the transportation point-of-view, the IoT has the potential to provide outstanding efficiency, which is leading humankind into the future of autonomous vehicles. Therefore, automated driving technology

[a] Center for Digital Safety & Security, Austrian Institute of Technology.
[b] Faculty of Computer Science, University of Vienna.
* Corresponding author: a.m20488@gmail.com

is considered the most vivid usage example of IoT [265]. According to Gartner, Inc.[1], it is expected that the enterprise and automotive IoT market will increase to 5.8 billion at the end of 2020, a 21% rise from 2019 [104].

Modern vehicles are considered IoT devices that contain intelligent units to facilitate driving. New cars contain smart devices that control the internal parts of the vehicle, which are responsible for producing motion. Nowadays, a new automobile contains over a hundred Electronic Control Units (ECUs) and has an internet connection to provide additional facilities [197]. The luxury car has over 100 million lines of code; regarding the rapid change in software development, this number is anticipated to increase. The software runs on hundreds of ECUs that communicate through different types of communication protocols and buses such as the Controller Area Network (CAN Bus), FlexRay, and Ethernet [66]. Modern vehicles are considered part of a more widespread ecosystem, including other traffic stakeholders, infrastructure, customers, and authorities. For smart public mobility, self-driving technology requires advanced approaches, which interconnect vehicles together and with other infrastructure units. Novel approaches to autonomous vehicles are sufficient for highway or country roads; however, these approaches are not ready for urban environments [311].

Safety concerns are one of the most critical engineering aspects in the automotive industry [204]. The vehicular safety systems can be classified into two main categories. The first category aims to minimize the severity of injuries after the accident, this category is called "Passive". The second one is called "Active", and is used to restrict accidents of vehicles. Car seat belts, airbags, windshields, and others are considered the best example of the passive category, where anti-locking brakes are an excellent example of the active one. The active-safety is able to identify different critical situations that lead to accidents and force corrective actions to prevent accidents [356].

In the vehicular industry, safety concerns are increased regarding the replacement of the mechanical units with embedded systems [237]. The integration of the internet connection with modern vehicles to provide advanced features such as autonomous driving, communicating with cars or anything, software updating, and others, opens a new relevant topic in the vehicular industry which is cybersecurity. Any device connected to the internet can be exposed to multiple kinds of malicious attacks, and the same is true for the modern vehicle [246]. A vulnerable point in a car could lead to the entire vehicle being out of control. Attack could happen in case malicious code is injected into the communication bus in a vehicle. That could affect the standard functionality of the vehicle and cause tragic consequences. Moreover, it is necessary to study the exact security vulnerabilities in the initial phases of the vehicle engineering process because

---

[1]https://www.gartner.com/en.

afterward, it becomes cost-ineffective and challenging to add security counter-measures [197].

This chapter is organized as follows; Section 10.2 gives an overview of the safety and security in the automotive domain and discusses the different levels of automation briefly. The approach is defined in Section 10.3; then Section 10.4, describes the main structure of the proposed ontology-based security framework. Section 10.5 illustrates the other application domains that the proposed model can be used for to perform the verification and validation process of the selected security requirments against the existing potential security weaknesses.

10.2 Overview

Security and safety engineering are strongly related disciplines that benefit significantly from each other if the interaction is described sufficiently. These domains should be incorporated from the early stages and beyond in the system development process. Both safety and security are the main characteristics of system engineering. Furthermore, cybersecurity and functional safety need systematic methods for the current and future automotive industry [237]. These issues are important not just in the automotive industry but in all IoT systems. This section gives an overview of the functional safety and security in the automotive domain.

10.2.1 Safety in the automotive domain

Safety engineering is used to create safe systems with a focus on reliable parts [313]. The vehicular domain contains a collection of developed safety standards which are used to assure that all safety risks are decreased to a tolerable level. In addition, these standards are applied to design systems and safety-critical components [312]. Faults or deficiencies in either reliability for safety or computer-based automotive systems can cause a severe impact. At least one death was traced back to the defects of the automotive computer system. Roughly 500 injuries and fatalities were allegedly attributed to some faulty vehicle designs by the same Original Equipment Manufacturer (OEM) [203].

The ISO 26262 [175] is a worldwide accepted standard for automotive electrical and/or electronic systems design and development [111]. It is the first complete standard to address the functional safety of the Electrical/Electronic (E/E) systems in the current automotive domain [9]. The standard is a framework that combines functional safety into the development life cycle of the automotive industry [111]. By the ISO 26262 safety standard, the lifecycle of an automotive component begins with determining where the system is being used and how important it is for automotive safety [341]. The standard sets out four levels of potential safety impact, known as Automotive Safety Integrity Level (ASIL). ASIL is one of the four levels (e.g., A, B, C, and D) to identify safety measures and requirements which are required by the system to prevent unreasonable residual risk. Level D is defined as the highest level, where A is the lowest one [178].

The American Automobile Association (AAA)[2] performed tests on cars with pedestrian detection alerts and automatic emergency braking using dummy pedestrians. The automobiles hit the dummies in 60% of the trials. The result became much worse, at 89% when the researchers replaced the adult pedestrian dummies with a kid version [19]. Modern pedestrian detection systems can inform the driver when there is a serious risk of a collision through an audible, visual, or haptic signal. Roughly 56% of 2018 car models are fitted with automatic emergency braking and pedestrian detection capability according to the research report by AAA [4].

10.2.2 *Autonomous vehicles*

According to the National Highway Traffic Safety Administration (NHTSA)[3], automated vehicles will help to reduce the injury rate and save lives, as 94% of serious accidents are human mistakes [266]. The Society of Automotive Engineers (SAE)[4], defines six levels of automation in the range between (0) to (5), where the level (0) is non-driving automation, and (5) is full driving automation [304]. As discussed in [340], the six levels of automation are:

- **Level 0 (No Driving Automation):** driving is performed manually. The driver offers the "driving mission," while systems assist the driver.

- **Level 1 (Driver Assistance):** this is considered the lowest level of automation. The vehicle has only one automatic driver assistance system, such as accelerating or steering.

- **Level 2 (Partial Driving Automation):** the level of automation includes Advanced Driver Support Services (ADAS) that is considered a short self-driving because a driver always takes over the control of the vehicle from time to time. The best example of level 2 automation, for instance, is Tesla Autopilot and Cadillac Super Cruise.

- **Level 3 (Conditional Driving Automation):** the vehicles of level 3 are equipped with "environmental detection" and can make decisions on their own, but they still necessitate watchful driver.

---

[2]https://www.aaa.com/International/.
[3]https://www.nhtsa.gov.
[4]https://www.sae.org.

■ **Level 4 (High Driving Automation):** the vehicles with level 4 can intervene when something goes incorrect, or system failure occurs. These vehicles work in self-driving mode but in a defined environment.

■ **Level 5 (Full Driving Automation):** this level requires no driver attention. The vehicles with level 5 can interact as an experienced driver can do; they can move anywhere without geofencing.

According to the World Bank, in the developing economies, the traffic jam can cost up to 5%, and 0.5–3% for developed economies, of their annual Gross Domestic Product (GDP). The traffic congestion affects passenger timetables and also reduce the global economy to $1.4 trillion per year [64].

10.2.3 Cybersecurity in the automotive domain

All road transport actors, road infrastructure, and authorities should cooperate in developing fully or highly self-managed vehicles. Their parts are influenced by an integrated infrastructure system, which requires new reliable communication approaches that enable vehicle-to-vehicle, and vehicle-to-infrastructure communication. Therefore, the primary requirement for processing different vehicle states and accelerating further growth is ensuring a stable, reliable, and secure connectivity. Reliable communication is considered the critical requirement for processing and speeding up the development of different motor vehicle systems [311].

In 2010 the automotive sector started to focus more on cybersecurity when research groups from Washington University and California San Diego demonstrated the ability to control critical components in a vehicle such as a brake system by injecting a code into the CAN bus. The following year, three different ways of remote attacks were shown by the same group for executing the code on a vehicle, including the mp3 radio, Bluetooth, and telematics unit, when the code injected into the CAN bus could impact any physical unit in it. The study from remote attacks is noteworthy because it revealed that cars were not only locally vulnerable but also could be attacked from across the world [254]. The authors in [254] have collected data on many vehicles' architecture to classify which vehicles can pose the lowest challenges to an intruder. Finally, they found that the 2014 Jeep Cherokee attack looks as though it combined broad attack surfaces, simpler architecture, and several advanced physical features that would be used in their research. The same work carried out a remote attack on a 2014 Jeep Cherokee and similar cars which physically influenced some parts of the vehicle.

In the automotive sector, there are hundreds of units in a vehicle; each unit can have potential security gaps that need to be examined to understand the risk and how to address a particular risk with applicable security mitigations. Connecting a physical device to the internet means new cyber risks can be generated. The following sections address the structure of the proposed ontology-model for the vehicle sector for performing security requirements verification & validation.

10.3   Approach

Currently, the automobile sector is not protected by a domain-specific risk management framework. A common working group has been founded by the International Organization for Standardization (ISO) and SAE, establishing a standard for cybersecurity engineering of road vehicles (ISO/SAE 21434) [176], which is going to be published in 2020 [311]. Determining whether a system is entirely secure is very difficult. A system may be fully operational but can be vulnerable due to other unforeseen tasks during the process [197].

Modern vehicles are considered sophisticated systems that include numerous electrical units and communication protocols. Consequently, cybersecurity in the vehicular industry is quite challenging. Any vulnerable unit in a vehicle may lead to complete attacks on the whole vehicle. New methodologies should be created that can manipulate an enormous amount of vehicle data to ensure a high level of protection in modern and future vehicles. These methodologies help to find the most relevant security requirements for the development of a secure vehicle. The complexity lies in manipulating data, where a large amount of data has to be appropriately managed, processed, and manipulated. The data can be derived from multiple dimensions, such as the following:

- **Assets:** assets in a vehicle are known as information, machinery, an element or a physical object or a logical object [6]. The modern vehicles contain a wide range of ambiguous units and drivers of communication protocols.

- **Potential Threats:** Threats refer to the potential vulnerabilities which may be exploited by attackers [292]. During the risk assessment, security threats must be viewed in the course of a threat analysis. Where these threats have been detected, a vulnerability analysis should be performed to specify security requirements [315]. In the vehicular domain, the threat analysis will play an integral role in identifying potential negative actions affecting the vehicle security. The threat analysis method can be divided into the following fundamental actions [314]:

    1.   Design the vehicle components with all the necessary information and assumptions related to security

    2.   Model potential opponents with their qualifications, actions, tactics, techniques, and procedures

3. To identify potential threats, apply the risk model to the process model, which is defined in step (1) and (2)

4. Assess all threats which are detected to define the exact severity risk level

5. Select the security countermeasures to update the system design, to minimize the risk

6. To detect missed or new threats, repeat step (3)

■ **Vulnerabilities:** Security weaknesses are always present in computing and network systems. The attacker aims to manipulate the existing weaknesses to make it easier to access the system [271].

■ **Exploits:** Finding and manipulating security weaknesses to access systems is both an art and science. It is considered a game between security professionals and attackers [271]. Attackers exploit vulnerabilities for malicious activity [292].

■ **Security Requirements:** In the system development process, choosing security countermeasures should be seen as a significant task in order to keep the overall risk at an appropriate level of security. Furthermore, the identified potential threats and security vulnerabilities must be addressed by applicable security requirements to reduce the impact of the identified risk from an extreme or a high level such as an unacceptable risk level into a low-risk level which is acceptable with low impacts [6].

All points that are mentioned above contain a vast amount of data that needs to be managed smartly, and accurately to avoid missing security flaws in the vehicular development life-cycle. Furthermore, the authors propose to use the ontology approach to manage all the data to create complete relationships between all points in the vehicle such as assets, threats, vulnerabilities, and security requirements to validate the chosen security requirements against the defined threats. The validation process plays an essential role in the vehicle development life-cycle to determine the specific security gaps in the vehicle after applying security requirements and to define which points still need more protection. This work is based on the continuously changing the simulation model for performing the validation process over a considerable amount of vehicular data in an ontological form. Furthermore, all vehicle data will be represented in classes, subclasses, individuals, properties, and annotations of all assets, detected potential threats, detected vulnerabilities, and the selected security requirements with all related security properties. The ontology describes a set of primitive members in order to design a knowledge domain. The primitives for expression typically are class attributes or relationships. They include information on their meaning and the logical constraints of their application [310]. Ontology is a powerful tool used with standard knowledge representation such as terminology, taxonomy, groups, individuals, and annotations [322]. The function of an ontology is to act like a human brain. It operates with thoughts and relationships between several entities. This is seen as the way people perceive interconnected concepts [277]. The implementation of hundreds or thousands of requirements is a difficult task, as it takes a long time and is complex. In reducing query complexity, the architecture of the ontologies plays a significant role [75].

10.4 The structure of the ontology-based security framework

This section discusses the structure of the proposed ontology security framework. The framework is a simulation that intends to introduce a model for security requirements for the current and future vehicle domain. The framework evaluates the security level that is achieved after the security requirements have been applied to the vehicle, and then it provides a new range of security requirements that could be applied to the vehicle to improve that level and meet the actual security goal.

Security testing follows a series of activities that use objects aimed at preparing, planning, defining, executing, and reviewing test results and revisiting preparation [309]. Automotive security testing requires a robust method to manage all vehicle security details and to manage a wide range of security requirements to ensure a vehicle's security level. Therefore, the proposed model aims to automate the validation process of security requirements to conduct the validation theory by simulation. System simulation is the operation of a model or simulator representing the system. The model is configurable, which would be impossible, too costly, or unusual to perform on the entity it depicts [264]. According to to [330], there are mainly three types of simulation model:

- **Continuous Change Model:** this model is defined by a collection of continuous parameters, continually evolving according to a set of differential equations. Continuous models of transition are ideal for describing motions of physical systems by Newtonian or classical electromechanical forces. It also is used to represent manufacturers' activities.

- **Continuous Change Model:** In this type of model, time is broken up into a series of finite cycles, and model variables can only change at the end of cycles. The way they adjust is defined by a set of differential equations that relate to values in the present time period to those in the preceding time period. This model was mostly used in Econometrics, where data is only available periodically.

■ **Discrete Event Model:** the parameters of this model are discrete quantities that describe the states of the objects in the system. Interactions among objects occur only at certain times, divided by inactivity periods. These interactions are often referred to as "events". This form of simulation refers to a wide range of movement unit systems such as neutron flow in nuclear reactors, queueing, and customer congestion at service points, transport systems, and others.

As is will be discussed later, the proposed security architecture is described as a continuous change model because it is focused on adjusting vehicle security parameter values, then identifying security vulnerabilities after and before it occurs regardless of time. Figure 10.1 illustrates the six main blocks of the proposed ontology-based simulation security framework.

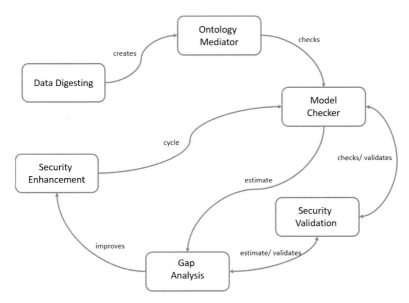

Figure 10.1: The building blocks of the simulation-based security framework.

10.4.1 Data digesting

The BigML[5] discusses comparing (big) data with the digestive system. There are many phases through which the data flows; some of these get too much focus while others need more [188]. These phases such as Ingestion, Digestion, Absorption, and Assimilation are discussed in [188]. The pre-processing data phases are proposed to be combined with this work to obtain the necessary knowledge

[5]https://bigml.com.

needed in the verification and validation process to validate vehicle security criteria against specific security vulnerabilities. These phases are described as follows:

- ■ **Ingestion:** data collection as follows:

  - ■ a list of all relevant details such as name, type, security mitigation on the components and assets of a vehicle,

  - ■ all threats listed with all related information typically name, Id, type, classification (category) and level of risk,

  - ■ all related security requirements with the identified security weaknesses in the system model.

- ■ **Digestion:** Processes original data in a formal language, which allows different values to be derived from the raw format.

- ■ **Absorption:** extracts all the data values from the input needed to create an ontological representation.

- ■ **Assimilation:** acts as a buffer to eliminate any unnecessary information. For example, threats with a low level of risk do not seem to be major security risks to a vehicle.

10.4.2 Ontology mediator

The second phase of the proposed framework aims to create a semantic description of the vehicular data. The data is presented in sets of vocabularies, taxonomies, and ontologies. It contains a set of defined terms which are essential to how information is expressed [159].

10.4.2.1 Vocabulary

It is a simple, well-defined set of terms with definite meaning in all contexts [159]. Vocabulary is used to identify the terms used for a specific function, to explain reasonable relationships, and to establish relevant constraints using these terms [367].

10.4.2.2 Taxonomy

The Ref. [52] defines multi-layers of a security framework to provide the appropriate level of protection to vehicles. This approach implements a defense technique, believing hackers can gain access across individual layers [52]. To fix the structural ambiguity of hundreds of internals units in vehicles the taxonomies in this proposed work describe the terminologies of vehicles according to a set of classes and subclasses as multi-layers. This helps in defining several object properties that define the relationships between different entities in the ontology structure. The taxonomy is designed to semantically identify all data in the vehicle that will be used during the validation process, as will be discussed later. This design contains a hierarchy of data knowledge of classes and subclasses representing the main entities which are required in the security validation process (as is depicted in Fig. 10.4, and will be discussed later in Section 10.4.2.3). These classes define vehicle data in related sub-categories of classification schemes.

Figure 10.2 illustrates the vehicular component layers. The component class classifies vehicle data into six categories; each category accommodates vehicular components related to the matched type. Also, the risk class contains four risk levels (low, medium, high, and extreme) that define the risk level of the existing potential threats. These levels measure a specific threat's severity level. Estimating the risk levels is necessary to determine the level of security requirements which are needed to address the identified threats. Vehicle sub-components schema look like separate component category layers; each layer contains components with common component types. In addition, each layer includes four sub-layers of risk levels to define the exact level of security related to each defined threat related to vehicle components.

This taxonomic structure gives a better understanding of the ontological hierarchy (vehicular units) that defines all the ontology individuals in one main class. However, from the algorithm performance point of view, not much difference is noticed between identifying all entities as individuals to similar classes (Components) or assigning separate subclasses. Figure 10.3 demonstrates the retrieval

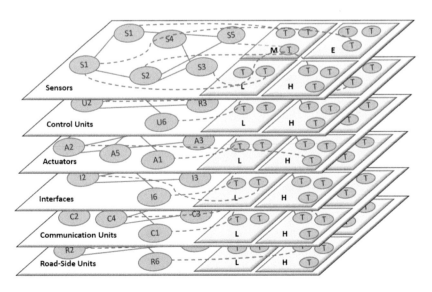

Figure 10.2: The taxonomy of vehicle components.

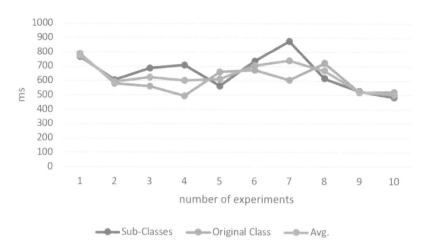

Figure 10.3: Querying performance according to the specified taxonomy.

time of individuals in one main class or subclass. This test is performed using Semantic Query-Enhanced Web Rule Language (SQWRL). SQWRL uses a default Semantic Web Rule Language (SWRL)[6] rule and uses it as a template specification for a query [273]. Nevertheless, other studies use SPARQL[7] language to test performance in the querying process. There is a notable difference in SPARQL's performance compared to SQWRL, where SPARQL is faster than SQWRL. In SPARQL, the inferred model is calculated one time only before the matching begins. Therefore, during the SPARQL execution time, the inference time is not considered. In the SQWRL scenario, the inference time influences the execution time [39].

10.4.2.3   Ontology

The ontology approach is considered the pillar of this proposed model to create a series of essential definitions of the main problem and related solutions which are able to tackle the security weaknesses in the vehicular domain [322]. The ontology determines a collection of primitive members to model a knowledge domain. The primitives of representation usually are classes, attributes, or relations. These include information on their meaning and restrictions on their consistent logical application. An ontology, therefore, represents the knowledge of a particular domain [310]. The security framework establishes multiple classes, categories, entities, properties, and annotations for all vehicle parts to be used in the verification and validation process. The ontology aims to explore the connections between vehicles, equipment, possible threats, and protections which are capable of addressing and anticipating actual security vulnerabilities.

---

[6]https://www.w3.org/Submission/SWRL/.
[7]https://www.w3.org/TR/rdf-sparql-query.

Ontology is an aspect of semantic technology, as being part of the W3C[8] semantic web standard stack [277]. The semantic web offers a robust and practical approach to the multiplicity of information and information services. In order to make the World Wide Web understandable, semantic web technologies are created. Web resources are distributed inherently, and the descriptions of resources contained on the Semantic Web are thus also distributed. The ontology is made up of statements describing definitions, relationships, and limitations. It is similar to a schema in a database or a hierarchy diagram of object orientation. For areas such as finance and medicine, an ontology can capture depth for representing similar objects. An effective ontology facilitates cooperation between applications in the sense of ontology [159]. The data in ontology is based primarily on a compilation of three complementary languages: the Resource Description Framework (RDF), RDF Schema (RDFS), and the Web Ontology Language (WOL) [159]. RDF presents a way to model knowledge but does not explain what it means. RDFS includes a particular vocabulary for RDF to describe taxonomies of properties, classes, domains, and ranges for context requirements; Where the WOL presents a vocabulary that expresses ontologies that encompass the domain knowledge semantically. Knowledge is expressed as a series of assertions/statements composed of three sections: subject, predicate, and object. Those three sections sometimes mean that sentences are often pointed to as triples. The three components of a sentence have meanings that, in plain English grammar, are identical to their meanings [159].

The core structure of the ontology design in this work is depicted in Fig. 10.4. The figure shows how the ontology entities are interacting together using the well-described entities and object properties. Some of the entities are defined as classes and subclasses for a clear understanding of ontology design, such as the relationship between components (class) and the assets (subclass).

For example, the vehicular units contain a set of critical assets that could be the real target of a cyberattack, such as "Cryptographic Keys" asset of the V2X HSM (Vehicle-to-anything Hardware Security Module) component as discussed in [63] by "Car 2 Car Communication Consortium"[9]. The relationship between the component and asset entities is defined as subClassOf.

The property of rdfs: subClassOf is the simplest way to define a class's membership and other related classes [159]. So the asset unit is a subClassOf the component unit, where the asset is a specialization of the component. There are many elements in a vehicle that must be validated entirely irrespective of any related assets. In this case, a particular property is established between the component and the individual assets. According to the above example, the class "Components" has an individual "V2X HSM" which is defined as an instance of this

---

[8]https://www.w3.org.
[9]https://www.car-2-car.org.

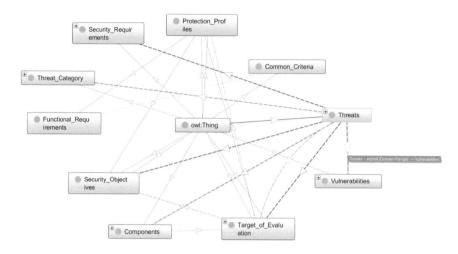

Figure 10.4: Ontology graph of the classes and subclasses of the proposed security framework.

class. The "itContains" property is inserted into both individuals in order to determine the connections as "V2X HSM" itContains "Cryptographic Keys". The connection between the "V2X HSM" as a component and the "Cryptographic Keys" is described according to a domain and a range relationship.

The domain sets the combination of individuals which are subject (resource) to the statement using the described property, where the range defines the combination of individuals which are object (properties) to the statement using the described property [159]. The components and assets comprise several protective measures that could be exploited by potential threats to attack the vehicle. The authors have identified new classes in the ontology taxonomy framework. The first is a class of "Properties" that contains the most common security characteristics to be defined as in-vehicle components and assets. The second class, the "Threats," includes a range of potential threats that could exploit vehicle components and assets' security vulnerabilities. Additional properties (predicates) are established to define communications between components/assets and properties/threats. Figure 10.5 shows how the interactions between components, threats, and properties look like.

The new property "hasVul" is asserted into component/asset (domain) and to the class "Properties" (range) to define the relationships between these two classes. In addition, "exploitedBy" property is defined between "Properties" (domain) and "Threats" (range). The affected vehicular unit has a property "attack" with the threat that exploited the security weaknesses.

In this study, the triples are specified to explain the relations between different entities in this proposed security framework. Multiple object properties are de-

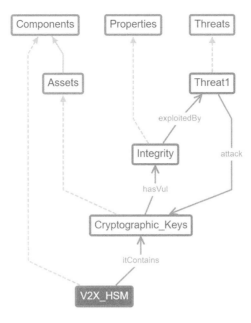

Figure 10.5: A Part of the ontology taxonomy of the components, assets, properties, and threats.

fined to explain the relationship between entities in the ontology. Here is a sample of established properties, as shown below:

1. $T \xrightarrow{\text{addressedBy}} SR$

2. $C \xrightarrow{\text{protectedBy}} SR$

3. $T \xrightarrow{\text{hasRisk}} R$

Throughout the ontology architecture, these statements are used to describe relationships between domain and range entities. For emphasis, the following formulas explain the formal description of these statements:

■ $T \xrightarrow{\text{addressedBy}} SR$: "Any threat needs to be addressed by a security requirement(s)"

   ■ $\forall_t \in D_{\text{Threat}}$, Threat (t) : t belongs to the domain "Threat".

   ■ $\forall_{sr} \in D_{\text{SecurityRequirement}}$, SecurityReq$(sr)$: sr belongs to the domain "Security_Requirement"

   ■ Threat(t): "t is threat"

   ■ Security_Requirement(sr): "sr is security requirement"

- addressedBy(t, sr): "t needs to be addressed by a sr"

- $\forall_t, \exists_{\geq sr}$ addressedBy (t, sr) "All threats must be addressed by one or more security requirements"

■ $C \xrightarrow{\text{protecteddBy}} SR$: "Any component must be protected by a security requirement(s)"

- $\forall_c \in D_{\text{Component}}$,Component (c) : c belongs to the domain "Component".

- $\forall_{sr} \in D_{\text{SecurityRequirement}}$,SecurityReq$(sr)$: sr belongs to the domain "Security_Requirement"

- Component(c): "c is component"

- Security_Requirement(sr): "sr is security requirement"

- protectedBy(c, sr): "c must be protected by a sr"

- $\forall_c, \exists_{sr}$ protected (c, sr) "All components has have to be protected by the existing security requirements"

■ $T \xrightarrow{\text{hasRisk}} R$: "Any threats have at least one risk level"

- $\forall_T \in D_{\text{Threat}}$,Threat (T) : t belongs to the domain "Threat".

- $\forall_r \in D_{\text{Risk}}$,Risk$(sr)$: r belongs to the domain "Risk"

- Threat(t): "t is threat"

- Risk: "r is a risk"

- hasRisk(t, r): "t has r"

- $\forall_t, \exists_{\geq r}$ hasRisk (t, r) "any identified threats must have at least one risk level."

10.4.3 Model checker

The model checker behaves as a peer assessment to check almost every node in the ontology design to verify the formal correctness of all vehicle entities of the ontological format. The model aims to check the consistency of the ontology design only once before starting performing the validation process. Each potential threat must have a relationship with at least one component with related security properties that have been exploited by it. This phase uses SQWRL language for performing the verification process and for checking the relation among different entities in the ontology. For instance, **"Any component affected by a threat needs to be protected by a security requirement"**, this statement is translated into a series of mathematical operations performed on ontology to test the coherence (true or false) of entities and relationships:

- $\forall_T \in D_{\text{Threat}},\text{Threat}\,(T)$

- $\forall_c \in D_{\text{Component}}\,,\,\text{Component}\,(c)$

- $\forall_{sr} \in D_{\text{SecurityReq}},\text{SecurityReq}(sr)$

- Component(c): c is component

- Threat(t): t is threat

- protectedBy(c, sr): "c needs to be protected by a sr"

- affectedBy(c, t): "c affected by threats"

- $\forall_c \{\exists_t [\,\text{affectedBy}\,(c,t)] \rightarrow [\exists_{sr}\,\text{protectedBy}(c,\,sr)]\,\}$

- $\forall_t \{\exists_{sr} [\,\text{affectedBy}\,(t,sr)] \rightarrow [\exists_{sr}\,\text{protectedBy}(c,\,sr)]\,\}$

To identify the security gaps in ontology, the verification process obtains results such as the mentioned query. The gap results are based on two fundamental parameter values, which we use in the absence of the automotive security standards which are based on IEC 62443 [171] and Common Criteria [174]. First, during the concept phase, the Security Target (ST) must identify the specific security goal to be accomplished. Security requirements are used to mitigate risk to an acceptable level. On the other hand, after implementing security requirements, the resulting state is called Security Achieved (SA). A way to manage the security requirements structure is by identifying them in categories known as protection profiles. The Protection profile (PP) is a document that outlines security and the resulting requirements for a particular Target of Evaluation (ToE) according to the Common Criteria (CC) [174]. The ToE is an abstract device or system unit definition for a particular purpose. The PP also describes the ToE(s)'s ST or security features. In order to develop secure vehicles, it is essential to ensure that one or more PP(s) comply with defined ToE(s). This is particularly important because vehicle systems are often re-used in a different context. Ensuring that such a system follows the PP in this sense means that protection needs are met [322]. In order to deal with the potential threats in the automotive field and keeping the risk low, this work uses the IEC 62443 standard series [171] to select relevant requirements. The main objective of IEC 62443 is to present a framework that addresses a wide range of security vulnerabilities in the Industrial Automation and Control System (IACS) [171]. The IEC 62443 security standard, defines the security requirements into four Security Levels (SLs), as discussed in [171]:

- SL1: Prevention of unauthorized data disclosure.

- SL2: Prevention of unauthorized data disclosure with low motivation and general skills.

- SL3: Prevention of unauthorized data disclosure with IACS distinct competencies and moderate motivation.

- SL4: Prevention of unauthorized data disclosure with IACS specific skills and powerful motivation.

10.4.4 Security validation

The validation process aims to determine whether the identified security requirements are validated against existing security risks. This process is a rule-based approach composed mainly of a series of "if-then" clauses. It creates SQWRL, which presents operators like SQL to obtain data from ontologies. Figure 10.6 illustrates the main phases of the proposed security requirements validation process.

A specific security requirement should tackle current potential threats found to attack a component or asset in the vehicle. The validation process uses the security characteristics of the predefined security requirements to update the affected component with new security characteristics. Then the rule engine uses components with modified security properties to check whether potential threats still exist or not. If no potential threats are created, the security properties which are used to update the component are validated to address potential threats.

Typically, rules engine support inferences beyond what description logic can deduce. The rules engine is powered by rules that are considered part of the overall

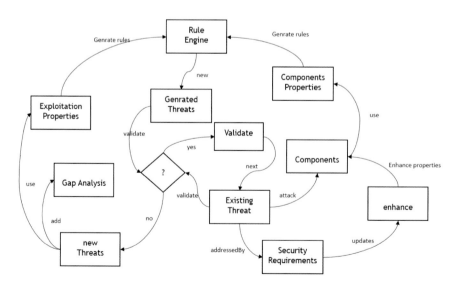

Figure 10.6: The main phases of the validation process.

representation of information [159]. Therefore, false findings are stored in the gap analysis table, which includes all the security vulnerabilities found in previous processes (verification and validation). The gap analysis approach aims to define the current level of security that is accomplished and deals with the existing security weaknesses by choosing other security requirements to deal with those weaknesses to increase the security level in a vehicle.

10.4.5   Gap analysis

The Gap Analysis phase is integrated with this work to evaluate the SA gaps before and after the verification and validation of the selected security requirements. It estimates if the identified security requirements are compatible with real ST and how the current security status could be enhanced. The method calculates the values of SA and ST, generating a full description outlining the definition of the impact on detected potential threats of the selected security requirements.

10.4.6   Security enhancement

This is the last step after all security vulnerabilities are well-defined; this phase aims to propose new security mechanisms that can resolve existing security risks. In this step, SWRL is used to reflect information that selects new security criteria to address potential threats and achieve the necessary ST. The following demonstrates a simple example of the SWRL declaration to identify a specific security requirement for a potential threat:

The reasoner uses the defined SWRL statement to recognize the specified statement's semantic meaning according to the structure of the ontology, as depicted in Fig. 10.4, and choose the proper security requirement that suits the statement clauses. The following outlines explain how the reasoner uses the statement of input and the ontology structure to infer the matched security requirement.

Reasoners add inference to the ontology design. Inference provides two additional logical structures as classification and realization. Reasoners apply inference to concept ontology. Classification and realization populate the structure of the class and makes it possible to connect concepts and relations to others appropriately. Reasoners typically are plugged into tools and frameworks such as Protégé[10] [159]. This work uses Protégé as an ontology editor and the Pellet as the reasoner. Pellet is an open-source, java-based OWL DL reasoning engine, supporting most of OWL's construscture [159]. The comprehension process which the reasoner follows to define the meaning of the predefined SWRL statement is described as follows:

---

[10]https://protege.stanford.edu.

1. *addressedBy Domain Threats* "The threats class is a domain of the addressedBy object property"

2. *Threat_1 addressedBy RE_1* "The security requirement RE_1 is already addressed the Threat_1"

3. *Threat_1 Type Spoofing* "The Threat_1 category is a spoofing according to the taxonomy of the aforementioned ontology"

4. *RE_2 subType "integrity"8sd:string* "The RE_2 contains a particular security property (that is integrity)"

5. *addressedBy Range Requirements* "The requirements class is a range of the addressedBy object property"

6. *Threat_1 subType "integrity"8sd:string* "The Threat_1 contains a specific security property (that is integrity)"

7. *Threat_1 addressedBy RE_2* "The RE_2 is inferred to address the Threat_1"

10.5 Application fields for the proposed model

The proposed ontology model can be used to verify and validate the security requirements against potential threats in a wide range of applications of IoT and Cyber-physical systems (CPS). The proposed framework aims to introduce a new methodology that can manipulate an enormous amount of vehicle data to ensure a high level of protection in future modern vehicles. This methodology helps in finding the most applicable security requirements that can be integrated with the vehicle to address particular potential threats. As the same in the Cyber-physical systems of production (CPPS) that consist of intelligent objects interacting on a global basis and exchange information to integrate and develop a wide range of existing technologies and modules such as robotics, industrial automation and control, IoT, big data, and cloud computing [235]. The IoT and services now contribute to a fourth industrial revolution in the manufacturing environment. Industries will create global networks, including CPS, equipment, warehousing systems, and production facilities. The CPSs include smart devices, storage systems, and plants that can share data autonomously trigger actions, and control one another independently in the manufacturing environment, smart devices, storage systems [191].

The complexity and variability of electronic parts within the IoT and CPS applications lead to increased security vulnerability. In addition, as mentioned earlier, the IEC 62443 standard is integrated with this work to select applicable security requirements that can be applied to the vehicular domain. This standard is essential to develop business security extensions that integrate the needs of IT systems with the particular requirements for powerful IACS [171]. The use of IEC 62443 into the proposed model also allows the model to be incorporated into the different design processes of smart applications, which verify and validate the security requirements that have been introduced to build a broad range of current and future IoT and CPS applications that become secure.

10.6 Conclusions

Modern vehicles have hundreds of electrical and electronic units for controlling and managing the critical safety components. The integration of internet technology with vehicles exposes them more to various cyberattacks. Accordingly, improving security in automobiles is essential to defend the vehicle from several cyberattacks. Traditional security verification and validation processes could miss some of the security vulnerabilities without proper security mitigations. Furthermore, this chapter introduces an ontology-based model that aims to create a knowledge representation of the vehicle details including components, assets, threats, vulnerabilities, and security requirements which are used in the verification and validation process. Then it defines a complete relationship among all vehicular points to validate the selected security requirements against the identified potential threats and the existing security vulnerabilities. This work is based on the continuously changing simulation model for presenting the validation method over a significant amount of vehicular details in an ontological structure. In addition, the model endeavors to give a new set of security mitigations able to improve the level of security to reach a satisfactory level of security protection. Finally, the chapter discussed that the proposed model could be applied in other application domains such as IoT and CPS to verify and validate the security requirements against potential threats.

Acknowledgment

This work has received funding from the iDev40 project. This project is co-funded by the consortium members and ECSEL Joint Undertaking under grant agreement No 783163. The JU receives support from the European Union's Horizon 2020 research and innovation programme, national grants from Austria, Belgium, Germany, Italy, Spain and Romania as well as the European Structural and Investement Funds.

Chapter 11

Dynamic Resource Provisioning using Cognitive Intelligent Networks based on Stochastic Markov Decision Process

Andrea Piroddi

11.1   Introduction

11.1.1   Problem domain

In all communication networks, resource management is a big issue. Channel establishment, bitrate adaptation [302] are just some of the most critical ones. In current communication systems the answer to these problems is normally provided by using complex and structured heuristic algorithms. This method focuses mainly on the identification of a mathematical model that can faithfully represent

University of the People, Italy.
Email: andrea.piroddi@uopeople.edu

the behavior of the analyzed system and on the resulting actions aimed at obtaining the desired behavior. The number of variables is very high, particularly in an urban environment where the structural layout and the dynamics of users do not help us; just think for example about bit error rate, bandwidth availability, spectral efficiency, interference, signal to noise ratio, power consumption, and more. On the other side we get the computational point of view in which the trend is exactly the opposite, that is, we try to simplify the transmission and receiving model as much as possible to reduce the effort in terms of computing resources. Another critical element that service providers must keep into account, while delivering interactive services, is to keep the tail of latency distribution as much short as possible at the growth of the network layout [47]. A simple way to slow down the variability of latency is to send the same packet to multiple receivers and use the result of the one that responds first, but this approach creates excessive redundancy within the system. Basically, we can say that network resource allocation can be seen as an optimization problem in which the node is an agent that acts in order to optimize its own ability to communicate to the other nodes by using an adequate number of resources. Using excessive resources is wasteful while using insufficient ones undermines the quality of the service [360]. In this hectic environment, Artificial Intelligence techniques are particularly suited if the problem domain shows the following characteristics [295]:

- the problem domain is constantly evolving as new devices, network nodes, links and others enter and leave the network following a stochastic distribution.

- the problem does not have a unique solution, but we can find several equally good solutions.

- there is no recipe book in which we can find algorithms that work well for every case.

- the problem is unbounded.

> This is the case of a communication network in a smart environment, where the agents perceive the state of the physical surroundings using sensors, and act on the system in order to optimize the specified performance measure [280].

11.1.2 Markov decision processes

For example, the receiving node is being interfered with and the transmitting node decides to increase the transmitting power level, or to modify the modulation, thus reducing the impact of interference. So, the agent transits in state s_{t+1} and receives the reward r_t. In our model, we consider both the transitions and

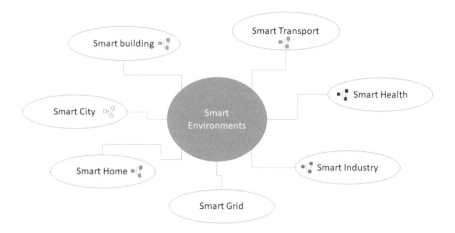

Figure 11.1: Smart environment.

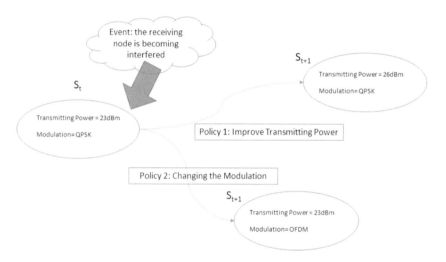

Figure 11.2: Reinforcement learning.

rewards as stochastic processes with Markovian properties, that is, the probabilities of the state transition and rewards depend only on the state s_t and on the action taken, a_t. It is essential to remember that in no case is the agent aware of what the status will be as a result of his action and what his reward might be. Only by interacting with the external environment during the training phase can our agent be aware of these quantities. In simple terms our agent (or network node) is going to learn from his own experience. We must therefore pass on to our agent the rules by which he can learn to do things in order to achieve the macro objective. Suppose we work with a finite set of states S and a finite set of actions A. We can define $P_0(.|s,a)$ which is the probability that the next state and the following reward belongs to the set U, where U is the set of S states per A actions. To complete the picture, we also can fix a discount factor γ, where γ belongs to the interval [0,1]. We know that a homogeneous finite-state Markov chain can be represented by a transition matrix A. For example, a 3x3 matrix represents a Markovian chain with 9 possible states.

$$A = \begin{bmatrix} 0.3 & 0.4 & 0.6 \\ 0.1 & 0.9 & 0.1 \\ 0.5 & 0.4 & 0 \end{bmatrix} \tag{11.1}$$

Taking into consideration the previous matrix, we can say that the elements of A represent the transition probabilities between the states of the chain, therefore a chain that is in the state i is likely to pass to the state j immediately in the following step, while the elements that stand on the diagonal of the matrix indicate the probabilities of remaining in the same state i. At this point we can define the transition state probability kernel, P, which for any (s,a,s') belonging to the triplet $SxAxS$ returns the probability that an agent has to transit from a state "s" to any other state "s'" provided that the action "a" is chosen when it is in the state "s":

$$P(s,a,s') = P_0(\{s'\} \times R|s,a) \tag{11.2}$$

Finally, we can add the immediate reward function to our model

$$r : S \times A \rightarrow R \tag{11.3}$$

that returns the expected immediate reward received by the agent when it is in the state "s" and takes an action "a" that is

$$r(s,a) = E[R_{(s,a)}] \tag{11.4}$$

What is the reward function for?

1. to stimulate the efficiency of decisions

2. to determine the degree of risk aversion of the agent

In particular, the set of states the agent can take at instant $(t + 1)$, that is S_{t+1}, is random so we have,

$$P(S_{t+1} = y \,|\, S_t = s, A_t = a) = P(s,a,y) \tag{11.5}$$

Also,

$$E[R_{t+1} \mid S_t, A_t] = r(S_t, A_t) \tag{11.6}$$

When the agent is in the state (t + 1), it will repeat the operation. The ultimate goal is to maximize the expected total discounted reward defined as:

$$R = \sum_{t=0}^{\infty} \gamma^t R_{t+1} \ \ with \ \ \gamma < 1 \tag{11.7}$$

Trying to give sense to this function in our specific domain, we can say that the state space is nothing more than the space of all the possible combinations in which a node can be, for example, distant from the next node, number of nodes to which it's connected, quality of the connections, quantity of data to be transmitted, and more, while the action space is composed of all those possible choices that the node can make to transmit the information packet to the destination node. The simplest approach that comes to mind to identify the best behavior that each node should implement when dealing with a given situation is to analyze all the possible choices and identify the one that leads to the maximization of our reward function. Obviously, this method is not suitable in a context like that of a communication network, since the multi-dimensionality of the state-action spaces does not make it computationally feasible. An approach that is computationally addressable is the one that envisages identifying value functions. That is, once calculated the optimal value function, allows us to identify an optimal behavior with relative simplicity. In a Markov Decision Model, every state (s) potentially reachable by the agent is associated with a recommended action (a). This solution is called Policy (π).

How to find the optimal policy? An optimal policy (π^*) is the sequence of actions that allows the agent to reach the goal with the maximum expected utility. It is the best road of all. To find it, the agent must be able to plan the moves in several shifts in advance. The time required to solve the problem affects the agent's decisions. The time horizon can be:

1. Finite. The agent must reach the goal within a set time T (or N moves). The solution must be found with a sequence of actions within the maximum time, otherwise it is useless. In this case the policy is not stationary because the optimal decisions in each state vary according to the time.

2. Infinite. The agent has no time limit to reach the goal or the maximum time T is very large. In this case the policy is stationary because it does not change with time. Actions depend only on the current state.

In a Smart Environment the first condition is applicable, in the sense that the agent has a time limit to reach the goal. Moreover, we must consider that, besides

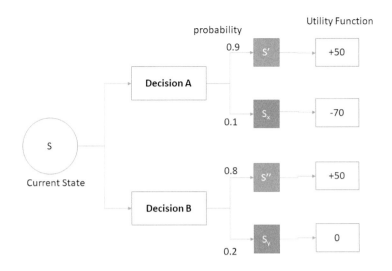

Figure 11.3: Decision policy example.

the probability of reaching the goal (S'), there is a risk of falling into a worse state (S_x). For example, BER calculation depends on the type of modulation and channel, which means that an inadequate choice of the type of modulation can lead to a need for retransmission of packets with a consequent additional cost or in other words for a negative reward. As shown in Fig. 11.3, before deciding, our agent must assess the consequences on the utility function of each possible future state. Decisions A and B offer the same improvement in terms of utility (+50). Decision A has a higher probability of success than B (90% against 80%). However, decision A is not the best if risks, for example an increase in interference, are also considered (−70 in 10% of cases). Decision B is the best option because it offers the highest expected utility (Expected Utility EU = 40), see Table 11.1.

In conclusion, in an MDP process the cognitive agent must consider both the probability of success and the risk of each action.

Table 11.1: Expected utility calculation example.

Decision A	*Probability P*	*Utility U*	*Expected Utility EU=U*P*
Result 1	0.9	50	45
Result 2	0.1	−70	−7
			38
Decision B			
Result 1	0.8	50	40
Result 2	0.2	0	4
			44

11.2 Reward function and the optimal policy

The algebraic sum of the rewards R obtained in the decision-making process measures the usefulness of the decisions.

$$U(S_0, S_1, ..., S_N) = \sum_{i=0}^{N} R(S_i) \qquad (11.8)$$

In this case we used the method of additive rewards. It is the simplest one because the reward in each individual state $R(s)$ is independent of time and position in the sequence of states. Alternatively, we can use the discounted rewards method. Each reward R is multiplied by a discount factor $\gamma \in [0, 1]$, raised to the position of the state in the decision sequence. In this case, the position of the state in the decision sequence modifies its reward.

$$U(S_0, S_1, ..., S_N) = \sum_{i=0}^{N} \gamma^i R(S_i) \qquad (11.9)$$

Therefore, if the rewards are discounted, the preferences are not stationary because the utility of the sequence changes by changing the order of the states. As we said in the previous paragraph, in a Markov decision model, every state s potentially reachable by the agent is associated with a recommended action (s). The chosen solution is called policy (π). Summarizing, we have

$$\begin{cases} T(s, a, s') \\ a = \pi(s) \end{cases} \qquad (11.10)$$

where T is the transition model, a is the action to perform, s is the current state, see Fig. 11.4, s' is the next state and π is the policy. If the policy provides the highest expected utility among the possible moves, it is called optimal policy $(\pi*)$. In this way the agent does not have to keep its previous choices in memory. To make a decision, it only needs to execute the policy associated with the current state $\pi(s)$, see Fig. 11.5.

It is not enough to calculate the best move in the next round because the agent risks becoming myopic. For example, the first move may seem the best but then lead the agent to a dead end. Only an overview allows the agent to identify the best route from the start. There are different planning techniques to find the optimal policy. For example, the algorithm of value iteration and the algorithm of policy iteration. Another algorithm is called asynchronous iteration of policies.

Bellman equation [49] expresses the value of the optimal solution of a mathematical optimization problem that can be translated in terms of dynamic programming, or decomposable into a sequence of concatenated sub problems. It is

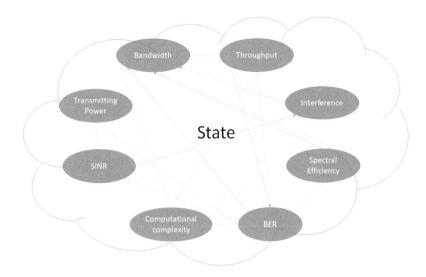

Figure 11.4: Determining elements of a state in a communication network.

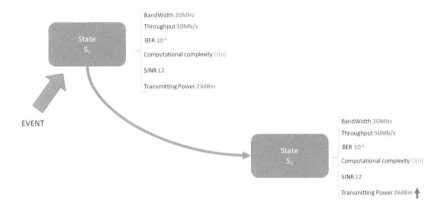

Figure 11.5: Example of state transition in a communication network.

natural to think that dynamism is induced by the passage of time, in many applications, such as that of a communication network, this is the real situation. The equation of Bellman exploits the so-called principle of optimality. This principle simply states that, if we consider the shortest route from New York to Toronto and it passes through Chicago, the Chicago-Toronto stretch of this optimal route is the shortest of all those traveling from Chicago to Toronto. The formalized version of the equation is as follows:

$$V(s_{t_0}) = \max_{a_{t_0}} \left\{ F((s_{t_0}), (a_{t_0})) + \beta V(s_{t_1}) \right\} \tag{11.11}$$

$$V^\pi(s_t) \equiv r_t + \gamma r_{t+1} + \gamma^2 r_{t+2} + \ldots \equiv \sum_{i=0}^{\infty} \gamma^i r_{t+i} \qquad (11.12)$$

We can observe that the sequence of rewards r_{t+i} is generated starting with state s_t and repeatedly uses the policy π to select the actions to be taken (that is, $a_t = \pi(s_t)$, $a_{t+1} = \pi(s_{t+1})$). Also, we have already seen the constant γ (with $0 \leq \gamma < 1$) is called discount factor and represents the interest of the agent with respect to the rewards he can receive "in the future":

1. if γ is close to 1 then the future rewards are relevant to the agent.

2. if γ is close to 0 then the future rewards are little relevant to the agent (for $\gamma = 0$, only the immediate reward is considered).

The quantity $V^\pi(s_t)$ is usually called a discounted cumulative reward. The final goal is to make the agent learn one policy π that maximizes $V^\pi(s)$ for all states s. This will be called an optimal policy and denoted as π^*.

$$\pi^* = arg\max_{\pi} V^\pi(s), \quad (\forall s) \qquad (11.13)$$

For simplicity, it will be referred to as $V^*(s)$, maximum discounted cumulative reward an agent gets starting from the state s. The cognitive node uses sensors in order to observe environmental changes and its neighbours' activities. The sensors could indicate a variation in the path losses, the presence of a new connected node, a change in the application which requires a different QoS needs, the activation of a new service. The cognitive agent must then identify the action, a, with which to respond. A fundamental element in RL algorithms is the propensity to put together the observation of the environment and the exploitation of data already obtained by the agent. Thrun [353] deals with this problem focusing on both the minimization of learning time (efficacy) and minimization of costs of learning (efficiency). A possible approach to build an efficient and effective agent in a smart environment is to make it use a strategy of direct exploration, so as not to waste the patrimony of events in which the agent itself participated. In [248] it has been pointed out that direct exploration approaches are superior to the indirect ones. We could say that history teaches.

11.3   Q-Learning technique

Reinforcement Learning is a kind of Machine Learning approach where a learning algorithm is trained on feedback rather on predefined data. These algorithms are publicized as the future of Machine Learning because these eliminate the cost of collecting and cleaning data. We're going to show how to implement a basic

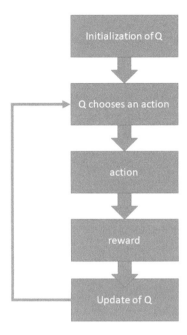

Figure 11.6: Example of a flowchart for a Q-learning algorithm.

Reinforcement Learning algorithm, called Q-Learning technique, to a communication network. As previously anticipated, the model envisages the introduction of an agent capable of observing the environment and carrying out actions; in this sense the agent/environment model can be modeled through a set of states S and a set of actions, by state, A. The model will, therefore, work as follows: by performing an action the agent will move from one state to another state and at each change of state a reward (a real or natural number) will be given to the agent, while the agent's goal will be to maximize the total reward. Therefore, the agent will continue to explore, learning the optimal actions associated with each state. Ultimately, the Q-Learning algorithm defines the way to calculate the quality of a state-action pair:

$$Q : S \times A \rightarrow R \tag{11.14}$$

In the initial phase Q will return a fixed value, defined at will during the initialization of the algorithm, then, each time the agent receives a reward (so at each change of status) new values will be recalculated for each state-action combination. The heart of the algorithm makes use of an iterative process of updating and correction based on the new information, see Fig. 11.6.

which translates into mathematical form and we have:

$$Q(s_t, a_t) \leftarrow Q(s_t, a_t) + \alpha_t(s_t, a_t) \times \left[R_{t+1} + \gamma \max_{a_{t+1}} Q(s_{t+1}, a_{t+1}) - Q(s_t, a_t)\right]$$

$$(11.15)$$

Where $\alpha_t(s_t, a_t)$ is the learning rate, R_{t+1} is the reward observed after running a_t in s_t, γ is the discount factor with $0 < \alpha \leq 1$, $\max_{a_{t+1}} Q(s_{t+1}, a_{t+1})$ is the maximum future value and $Q(s_t, a_t)$ is the old value. The formula above is equivalent to:

$$Q(s_t, a_t) \leftarrow Q(s_t, a_t)(1 - \alpha_t(s_t, a_t)) + \alpha_t(s_t, a_t)\left[R_{t+1} + \gamma \max_{a_{t+1}} Q(s_{t+1}, a_{t+1})\right]$$

$$(11.16)$$

An episode of the algorithm ends when the state is a final one (or state of absorption). Note that for all final states, s_f, $Q(s_f, a)$ is never updated and therefore retains its initial value. The learning rate determines with which extension the newly acquired information will overwrite the old information. A factor of 0 would prevent the agent from learning, and on the contrary a factor of 1 would cause the agent to only be interested in recent information. The discount factor defines the relevance of future rewards. A "zero" factor will make the agent "opportunistic", making it consider only the current rewards, while a factor tending to one will make the agent also attentive to the rewards he will receive in a long-term future. Let us imagine that the bot is a node in a communication network and is trying to transmit a data packet to one of its neighbors. He naturally concludes that the node to which the packet must be sent is within a maximum distance from the transmitting node. In addition, the destination node of the package has been identified appropriately and this can help the acting node in transmitting data securely. We consider a positive clue when the agent receives a confirmation that the *BER* on the channel is decreasing and a negative one when it increases. We may want to train our agent to improve the *BER* just acting on the power level or on the modulation scheme or on any other element it is able to adjust, noting that in this scenario we're not concerned about the latter. The bot is going to get the maximum reward when receives the confirmation of receipt of the data by the concurrent node. In Fig. 11.7, we can see an example of this setting.

By running the algorithm, we observe that its convergence occurs after about 200 iterations, see Fig. 11.8.

Before moving on, we must mention the curse of dimensionality. Consider a very large (or infinite) state space, as that of a communication network, in this case it is not possible to keep a value in memory for each state. When we are in this situation, we must try to identify an estimate of the values in the form

$$V_\theta(s) = \theta^T \phi(s) \quad s \in S$$

$$(11.17)$$

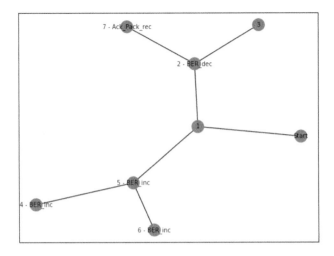

Figure 11.7: Example of graph structure.

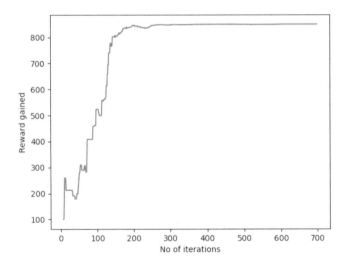

Figure 11.8: Reward gained.

where $\theta \in R^d$ is an array of parameters and $\phi : S \rightarrow R^d$ is a mapping of the set of states to a set of d-dimensional vectors. For the state s, the components $\phi_i(s)$ of the array $\phi(s)$ are named the features of state s and ϕ is named the feature extraction method. The set of ϕ_i represents the basis function set. It would be spontaneous to think that with a reasonably large set of training data, we could always approximate the expected value using the average value of k-nearest-neighboring values, since we should be able to find a consistent set of data and mediate between them. Unfortunately, our intuition collides with the harsh reality when the number of dimensions is very large. In fact, we assume a Hypercube of size q. Suppose we set a "reference" hypercube around a target point to capture a fraction r of the observations. This corresponds to taking a fraction r of the observations. It is as if we took a fraction r of the unit volume, into consideration that is, the expected length of the edge will be $e_q(r) = r^{1/q}$. In the case of 10 dimensions and $r = 0.01$, $e_{10}(0.01) = 0.01^{1/10} = 0.63$ and in the case of $r = 0.1, e_{10}(0.1) = 0.1^{1/10} = 0.80$ while the whole range of possibilities is upto 1.0. Basically, to capture 1% or 10% of the data to calculate a local average we must cover about 63% or 80% of the range of each input variable. Such variables are no longer local. Another consequence of scattered sampling in the case of high dimensionality is that all the samples are close to one side of the sample. Consider N points uniformly distributed in a q-dimensional space centered in the origin of the axes. The average distance from the origin to the nearest point is given by the equation:

$$d(q,N) = \left(1 - \frac{1}{2}^{1/N}\right)^{1/q} \tag{11.18}$$

For $N = 1000$ and $q = 20$, $d(q,N) = 0.69$, that is more than half the length of the side of the hypercube. So, most of the points are closer to the limit of the sample space than at any other point, and this complicates the calculation of the prediction.

11.4   Multi-Objective Reinforcement Learning (MORL)

Mezura [252] observes that in real-life many problems can be approached with multiple objectives function. For example, in a communication network some of the primary criteria to be considered are the latency and the energy consumption which are conflicting elements. The system engineer that needs to optimize more than one objective, must face a hard challenge because it is not always clear how different objectives influence each other. The optimization problems we have considered so far had only one goal, that is, to minimize (or maximize). From now on, we will consider the following multi-objective optimization problem:

$$\min_{x \in F}\{f_1(x), f_2(x), ..., f_k(x)\} \tag{11.19}$$

Definition 11.1   Dominance. Given two vectors $z^1, z^2 \in R^k$, we say that z^1 dominates z^2 according to Pareto $(z^1 \leq pz^2)$ when it gives,

$$z_i^1 \leq z_i^2 \ \forall i = 1,2,3,...,k \ and \ z_j^1 < z_j^2 \tag{11.20}$$

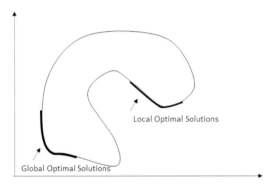

Figure 11.9: Pareto front.

for at least one index $j \in 1, 2, 3, ..., k$. Correspondingly we say that a vector of decisions $x \in F$ dominates another vector $x \in F$ and if it gives,

$$f(\hat{x}) \leq pf(x) \tag{11.21}$$

Definition 11.2   Pareto Optimality. A vector of decisions $x^* \in F$ is a Pareto Optimal if there is no other vector $x \in F$ that dominates it according to it, or if there is no other vector $x \in F$ such that:

$$f(x) \leq p \cdot f(x^*) \tag{11.22}$$

Correspondingly we say that a vector of objectives $z^* \in Z$ is optimal according to Pareto when there is no other vector $z \in Z$ such that

$$z \leq p \cdot z^* \tag{11.23}$$

So, if we are at an optimal point according to Pareto and we want to further decrease the value of one or more objective functions we must be willing to accept a consequent increase in some (or all) of the remaining functions of the problem. In this sense we can affirm that, in the space of objectives, the Pareto optimal solutions are points of equilibrium which are on the frontier of the set Z, see Fig. 11.9.

Based on the role played by the decision maker in the problem-solving strategy, the solution methods of multi-purpose programming are often divided into four broad categories:

1. **Method without preferences** in which the agent has no role and is considered satisfactory to have found any Pareto Optimum.

2. **A posteriori method** in which the set of all the Pareto Optimums is generated and then presented to the agent who chooses the best solution for him.

3. **A priori method** is one in which the agent specifies its preferences before the resolution process begins. Based on the information received from the agent, the best optimal solution is directly found, without having to generate all the Pareto Optimums.

4. **Interactive methods** are those in which the decision maker specifies its preferences as the algorithm proceeds, thus guiding the resolution process towards the solution that is most satisfactory for him.

The Q-Learning algorithm belongs to the second category, that is, posteriori method. The methods belonging to this class are also known as methods for generating the set of Pareto solutions. In fact, since the preferences of the agent are considered only at the end of the resolution process, all optimum points are generated according to Pareto. Once the set of Pareto solutions has been generated, it is presented to the agent which selects the best vectors for it. The main drawback of this strategy lies in the fact that the Pareto process of generation of the optimum is very often computationally burdensome. Furthermore, it may not be easy for the agent to choose a solution among the best ones presented to it, especially if they are numerous. For this reason, the way in which solutions are presented to the agent is very important. In this regard, we introduce the definition of the representation in the normal form of a game and the Nash equilibrium.

Definition 11.3 The representation in normal form of a game with n players, specifies the space of the strategies of players $S_1, S_2, ..., S_n$ and their reward functions $u_1, u_2, ..., u_n$. This game is indicated with $G = \{S_1, S_2, ..., S_n; u_1, u_2, ..., u_n\}$

Definition 11.4 In the normal game with n players, $G = \{S_1, S_2, ..., S_n; u_1, u_2, ..., u_n\}$, the strategies $s_1^*, s_2^*, ..., s_n^*)$ are in Nash equilibrium if, for each player i, s_i^* is the best response of player i to the specified strategies for the other $n-1$ players, $(s_1^*, s_2^*, ..., s_i^*, ..., s_n^*)$:

$$u_i(s_1^*, s_2^*, ..., s_{i-1}^*, s_i^*, s_{i+1}^*, ..., s_n^*) \geq u_i(s_1^*, s_2^*, ..., s_{i-1}^*, s_i^*, s_{i+1}^*, ..., s_n^*) \quad (11.24)$$

for every admissible strategy $s_i \in S_i$, that is s_i^* solves the equation:

$$\max_{s_i \in S_i} u_i(s_1^*, s_2^*, ..., s_{i-1}^*, s_i^*, s_{i+1}^*, ..., s_n^*) \quad (11.25)$$

The fundamental characteristics of our system are: (i) the actions of each agent occur in succession, (ii) all the previous actions are observed before the next action is chosen and finally (iii) the agents' rewards at each admissible combination of actions are common knowledge. The solution of such a scenario can be obtained by applying a backward induction procedure of the following type. Once the agent has taken action a_1, agent 2 is faced with the following problem:

$$\max_{a_2 \in A_2} u_2(a_1, a_2) \tag{11.26}$$

Assume that for each a_1 in A_1 the optimization problem of agent 2 has a unique solution, indicated by $R_2(a_1)$. This is the optimal response of agent 2 to the action of agent 1. Since both agent 1 and agent 2 can solve the problem of agent 2 we can trace the previous equation to the following expression:

$$\max_{a_1 \in A_1} (a_1, R_2(a_1)) \tag{11.27}$$

This technique is named backward induction [134] and can be applied to a Multi-objective scenario. That is, maximization of the reward is a function of the action of the first agent. The main problem is that agent 1 does not know the transition model $P(s'|s, a_1)$ which defines the probability of reaching the state s' from state s after implementing the action a_1; nor does it know the reward function $R_2(a_1)$ that specifies the reward for each state. The goal is to use the reward information to "learn" the expected utility $V^{\pi(s)} = E\left[\sum_{t=0}^{\infty} \gamma R(S_t)\right]$ associated with each non-terminal state. Estimating direct utility has the advantage of reducing the reinforcement learning issue to an inductive learning problem. Remember, however, that the utilities associated with states are not independent. In fact, the utility associated with each state is equal to its own reward added to the expected utility of the subsequent states. Basically, the utility values follow Bellman's equations for a fixed policy:

$$V_j^{\pi}(s) = R_{j+1}(s) + \gamma_j \sum_{s'} P(s'|s, a_j(s)) V_j^{\pi}(s') \tag{11.28}$$

This group of equations can be solved to obtain the utility function. Once the utility function has been obtained, the agent can extrapolate the optimal action for the next step and so on.

11.5 Single-policy MORL

One of the most popular approaches of RL on Multi-Objectives problems is to use single-policy algorithms with the goal to learn Pareto optimal solutions. Single-policy MORL programming employs Scalarization Functions. This method allows the reduction of the dimensionality of the multi-objective setting to a single scalar dimension.

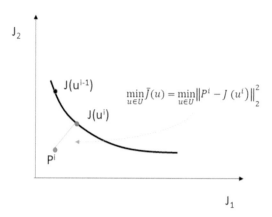

Figure 11.10: Reference point method.

Definition 11.5 Scalarization Function. A function can be defined as a scalarization function if it projects a vector $\vec{v}$ to a scalar:

$$v_w = f(\vec{v}, \vec{w}) \tag{11.29}$$

Where w is a weight array parameterizing f. A scalarization function can be introduced in different ways, although the most common one is the linear scalarization function. The weighted sum method is one of these implementations, achieving scalarization via convex combinations of single objectives using the weight array α:

$$\min_{u\in U}\vec{J}(u) = \min_{u\in U}\sum_{i=1}^{k}\alpha_i J_i(u) \tag{11.30}$$

although this is the most widely used scalarization method, it has strong limitations when working with non-convex problems, where it is impossible to calculate the entire Pareto set. A more complete approach is the reference point method, see Fig. 11.10, in which the algorithm must minimize the distance to an infeasible target point P with $P < J(u)$:

$$\min_{u\in U}\vec{J}(u) = \min_{u\in U}\|P - J_i(u)\| \tag{11.31}$$

11.6   Multi-policy MORL

Unlike the single-policy approach, multi-policy algorithms do not focus on reducing the dimensionality of the objective space but try to give a set of optimal solutions in one shot. Barrett and Narayanan [46] proposed to use the algorithm named CHVI (convex hull value-iteration) which, starting from the convex hull

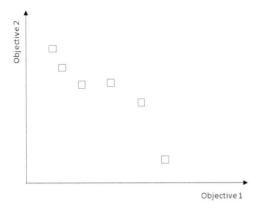

Figure 11.11: Pareto front of a bi-objective problem.

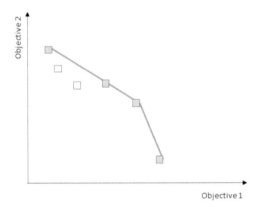

Figure 11.12: Convex hull of a bi-objective problem.

of the Pareto front, computes the deterministic stationary policies. The convex hull collects all the policies for which the linear combination of the value V^{π}, for the policy π, and some weight vector w [298] is maximal.

In Fig. 11.11 we can see the Pareto front of a bi-objective problem.

In Fig. 11.12 the blue line is the corresponding convex hull. Summarizing, the four yellow squares (deterministic policies) are the ones that Convex Hull value-iteration would give. The procedure of combining convex hulls is computationally expensive. In a recent study of Lizotte et al. (2010) they reduce the time complexity and the asymptotic space of the bootstrapping rule by making the agent give several value functions simultaneously. Moreover, they validate this method on clinical data for three objectives, although the possibility of applying this approach for higher dimensional spaces is not straightforward.

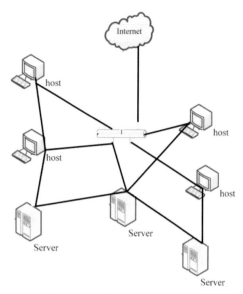

Figure 11.13: Sample network.

11.7 Case study—communication network subject to an interference attack

Let's try to contextualize these methods by applying them to a transmission network subject to an interference attack, a very common situation in an urban environment. This scenario is a rather frequent case of multi-objective and multi-policy analysis. The environment is defined as seven element-tuples:

$$Environment = (G, S, I, RI, P, R, \gamma) \tag{11.32}$$

The edges connect the nodes. An edge is able to run more than one service link, and the capacity of the service to answer requests (for example, *BW* available to that service) is represented by the β weight. The ability of a service to process a request (for example CPU time available for the service) is indicated by the weight c. These two values, β and c, are specific to the node, since the network is composed of low and high value nodes. Equations 11.33 and 11.34 formally describe the edges and nodes, respectively:

$$edges \subseteq \{(x, y) \mid x, y \in nodes, x \neq y\} \tag{11.33}$$

$$nodes \subseteq \{\beta, c \mid \beta \in N^{|services|}, c \in N^{|services|}\} \tag{11.34}$$

In our case study, the interference agent, at the initial stage, has an available number of resources to hit a device on the network. The quantity of resources is defined as $p \in N$ which it can use to interfere with one or multiple specific devices. In response, the interfered node can make available more resources to enhance the *SINR* to the compromised service. The match stops after a fixed number of turns. In an interference scenario, the engagements will persist until either the node establishes a good *SINR*, or the interferer ceases his attack. Anyway, this match focuses on what can be given by the node to attenuate the ongoing attack by using the resources it owns in an intelligent way. Thus, the definition of the state of the match is described by eq. 11.35:

$$\langle n, r \rangle \in S \qquad (11.35)$$

Where n and r are, respectively, the state of all nodes and the number of turns left in the match. The rules defined so far have the purpose of describing an interference scenario. The inability of the node to handle the job implies that the Tx/Rx action results are unavailable. Let's suppose that this approach represents a reasonable abstraction of an interference attack. The defender can only fight against interference increasing resources (power level, frequency hopping, modulation, etc...). The rules indicated below describe the interference action as zero-sum game. Our case study, uses the actions defined below, on a node, for reduction of interference (RI) and interferer (I):

$$RI = \{decrease\_interference, available\_resources, nothing\} \qquad (11.36)$$

$$I = \{increase\_interference, exhaust\_resources, nothing\} \qquad (11.37)$$

Let's see what the evolution of our system is: the system is in state $s_t \in S$, at a given time t. The algorithm checks if the system is at an end state. If not, it checks the interferer's available resources. If the interferer has power to inhibit a service, the agent can perform an action. If there are enough resources the attack can continue. The match stops when one of the following end conditions are met:

■ match stops if the interferer is able to place k adjacent interfering actions → Interferer wins

■ match stops if the interfered is able to place k adjacent interference reducing actions

■ match stops if its duration exceeds the number of established turns.

Taking an action: the interfered and interferer move one after the other. The service node can either allocate more power to the service or increasing resources. An action has a cost that is given by the weights β and c used in the system model. Once both agents have completed their actions for a turn, the reward is added to the cumulative total. The match goes on to the next turn until the end conditions have been met. The goals for the service node and interferer are the inverse of each other. The service agent aims to maximize the cumulative reward, as the interfering agent wants to minimize it. To define this reinforcement learning system, we need to be clear about the 3 main aspects:

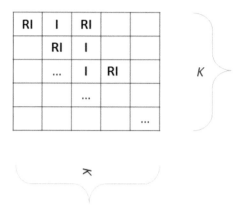

Figure 11.14: Blackboard.

- status

- action

- reward

The state considers the interference measured by the node. The action is represented by which positions a player can choose, based on the current state of the system. The reward is a value between zero and one and is allocated only at the end of the game. To update the state value estimate, we will apply the iteration of the value which is updated based on the following formula (see eq. 11.15):

$$V(S_t) \leftarrow V(S_t) + \alpha \left[V(S_{t+1}) - V(S_t) \right] \tag{11.38}$$

Having made the necessary premises, let us now consider implementing this logic in an interfered environment. Let's imagine our environment needs a State Class that our system will use as both blackboards, Fig. 11.14, and referees. The State Class will record the state of both the service node and the interferer and will update State if any agent takes an action. The referee is able to judge if the game is over and to give rewards to the agents accordingly. Suppose that, if the node manages to reduce the interference for $k \in N$ consecutive times (horizontal or vertical or oblique), it will be able to send or receive a packet. k is a value directly proportional to the quantization used in our transmissive network. Repeating this operation for the number of packets that constitute the data to be transmitted, we get to the completion of the service.

A Markov chain in a discrete time domain is a Markov process with a finite number of states. A Markov chain is specified in terms of its state probabilities

$$p_i[n] = P[X_n = a_i] \quad i = 1, 2, 3... \tag{11.39}$$

With the transition probabilities

$$\pi_{ij}[n_1, n_2] = P\{x_{n_2} = a_j | x_{n_1} = a_i\} \tag{11.40}$$

from theory we know that

$$\sum_j \pi_{ij}[n_1, n_2] = 1 \quad \sum_i p_i[k]\pi_{ij}[k, n] = p_j[n] \tag{11.41}$$

Moreover, if $n_1 < n_2 < n_3$, then

$$\pi_{ij}[n_1, n_3] = \sum_r \pi_{ir}[n_1, n_2]\pi_{rj}[n_2, n_3] \tag{11.42}$$

This is the discrete form of the Chapman-Kolmogorov equation [281]. It is possible to demonstrate that for a stationary, memory-free and ordered flow of requests, the number of points belonging to the τ interval is distributed according to a Poisson law of mathematical hope $a = \lambda$ where λ is the density of the flow (average number of events per time-unit). The probability that exactly m events occur in time τ is equal to

$$P_m(\tau) = \frac{(\lambda\tau)^m}{m!}e^{-\lambda\tau} \tag{11.43}$$

In particular, the probability that the interval is empty (no event takes place) is

$$P_0(\tau) = e^{-\lambda\tau} \tag{11.44}$$

The law of distribution of the amplitude of the interval between close events is an important characteristic of a flow. Consider the random variable T, which is the time interval between two arbitrary events close together in an elementary flow, and determine its distribution function

$$F(t) = P(T < t) \tag{11.45}$$

Going to the probability of the converse event, we get

$$1 - F(t) = P(T \geq t) \tag{11.46}$$

This is the probability that in a time interval of duration t, which begins at the instant t_k of the occurrence of one of the events, no subsequent event will occur. Since an elementary flow is without memory, the presence of any event at the beginning of the interval (at point t_k has no effect on the probabilities of the occurrence of subsequent events. For this reason the probability $P(T \geq t)$ can be calculated according to the formula

$$P_0(t) = e^{-\lambda\tau} \tag{11.47}$$

from which, we get,

$$F(t) = 1 - e^{-\lambda \tau} \quad (t > 0) \tag{11.48}$$

Deriving, we obtain the probability density as,

$$f(t) = \lambda e^{-\lambda \tau} \quad (t > 0) \tag{11.49}$$

Assuming that $a_1 = RI$ and $a_2 = I$, and that

$$P\{x(t + \Delta t) = RI \mid x(t) = I\} = 1 - \mu_1 \Delta t = \pi_{11}(\Delta t) \tag{11.50}$$

$$P\{x(t + \Delta t) = I \mid x(t) = I\} = 1 - \mu_2 \Delta t = \pi_{22}(\Delta t) \tag{11.51}$$

In this case, $\lambda_{12} = \mu_1$, $\lambda_{21} = \mu_2$

therefore, using the Kolmogorov equations, we obtain:

$$p_1'(t) + \mu_1 p_1(t) = \mu_2 p_2(t) \tag{11.52}$$

Since we know that $p_2(t) = 1 - p_1(t)$, we can conclude that

$$p_1(t) = \frac{\mu_2}{\mu_1 + \mu_2} \left[1 - e^{-(\mu_1 + \mu_2)t} \right] + p_1(0)e^{-(\mu_1 + \mu_2)t} \tag{11.53}$$

Observing that:

$$p_1(t) \xrightarrow{t \to \infty} \frac{\mu_2}{\mu_1 + \mu_2} = p_1 \tag{11.54}$$

$$p_2(t) \xrightarrow{t \to \infty} \frac{\mu_1}{\mu_1 + \mu_2} = p_2 \tag{11.55}$$

The transition probabilities are:

$$\pi_{11}'(\tau) + \mu_1 \pi_{11}(\tau) = \mu_2 \pi_{12}(\tau) \pi_{11}(0) = 1 \tag{11.56}$$

$$\pi_{22}'(\tau) + \mu_2 \pi_{22}(\tau) = \mu_1 \pi_{21}(\tau) \pi_{22}(0) = 1 \tag{11.57}$$

Where

$$\pi_{12}(\tau) = 1 - \pi_{11}(\tau) \pi_{21}(\tau) = 1 - \pi_{22}(\tau) \tag{11.58}$$

So, we obtain:

$$\pi_{11}(\tau) = p_1 + p_2 e^{-(\mu_1 + \mu_2)\tau} \tag{11.59}$$

$$\pi_{21}(\tau) = p_2 + p_1 e^{-(\mu_1 + \mu_2)\tau} \tag{11.60}$$

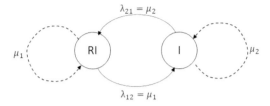

Figure 11.15: State transition probability.

These two equations provide us with the probability of staying in the *RI* state rather than going into the *I* state. Recall that if the node manages to reduce the interference for $k \in N$ consecutive times (horizontal or vertical or oblique), it will be able to send or receive a packet. This means that the probability to send a packet is given by:

$$P_{Tx\_packet} = \prod_{j=1}^{k} \pi_{11}(\tau + j) = \prod_{j=1}^{k} (p_1 + p_2 e^{-(\mu_1 + \mu_2)(\tau + j)}) \qquad (11.61)$$

We are therefore able to size our system in order to minimize the probability of waiting by the concerned node.

11.8 Conclusions

In this chapter we discussed a possible mathematical approach aimed at allowing a node (in a communication network) to optimize the use of the limited resources available. During this analysis we came across another fundamental question. How much information is needed by the node to make a good decision? By analyzing the context in which our node is, we were able to introduce the stationary Markovian processes, which depend only on the state of the node. We therefore followed the hypothesis that the parameters the node can vary to optimize resources are not unlimited, but are in finite numbers and often some of them are correlated. This implies that we can focus only on some of them, greatly simplifying the analysis of our problem. Therefore, the node must learn to manage the variation of these parameters to the best of its ability. In fact, each node will have different characteristics and attributes, in terms of computational and storage capacity, but the learning technique is the same for everyone. We can conclude that, the methodology linked to Reinforcement Learning based on Markovian processes allows the lightening of the computational aspect and makes it possible to extend this technique to all the nodes of the network.

Chapter 12

Data Model for Water Resource Management

Vlad-Andrei Nicolăescu,[a], Catalin Negru[a] and Florin Pop[b]*

12.1   Introduction

Water is a resource life depends on and managing it for local or national communities is becoming more and more challenging due to climate changes. The water quantity and quality are affected by many factors such as precipitation, drought severity and pollution. Approximately half of the world's population has difficulties in finding the water they need for at least one month every year, according to [250].

Today, satellite-generated images come to the aid of traditional, ground level techniques of measuring these parameters. However, datasets are dispersed across many platforms or websites, which makes planning the usage of water resources for people and industries harder. There are tools, such as the Water Evaluation and Planning (WEAP) system [327], which are useful in water resource management.

This tool is used in many papers that analyze the water needs of particular cities/areas throughout the world. A paper that uses WEAP to predict the wa-

[a] Computer Science Department, University Politehnica of Bucharest.

[b] Computer Science Department, University Politehnica of Bucharest, National Institute for Research and Development in Informatics (ICI), Bucharest.

* Corrresponding author: nicolaescu.vlad.ichb@gmail.com

ter needs for the Chinese city of Xiamen proves that, without additional water sources, there will be a shortage for the city by 2030 [205]. The paper analyses scenarios, trying to predict the water use for each type of industry (primary, secondary and tertiary), as well as household water used by citizens, taking into account the government's plans until 2050.

In arid areas where agriculture is supported by irrigation, the water levels of major lakes drop tremendously each year. A paper that analyzes the balance that must be met between agriculture and sustainable health of the environment in the Urmia Lake Basin area in Iran also uses the WEAP model [14]. This article analyzes 3 emission scenarios and 5 water management scenarios for the 2015-2040 period. This paper is representative for the study of how water resources in the basins of saline lakes should be managed in an efficient and sustainable way, both for human needs and the environment.

Another paper that uses WEAP aims to study the effects caused by both climate and socioeconomical changes on the usage of water resources [51]. This article uses an approach called "Decision-Making Under Uncertainty (DMUU)" for the Indian river basin of Cauvery. Socioeconomic changes usually have effects of unpredictable magnitude and are common especially in developing countries (such as India). The DMUU approach becomes admissible, also because of the ongoing climate changes. In the case of this paper, simulation is done for the 2021–2055 period.

The Water Management Data Model (WaM-DaM) [8] proposes a unification of all the heterogenous data sources in order to make water resource management more accessible. This model tries to merge the benefits of other tools and models (such as WEAP).

Motivated by the idea to bring together heterogenous data from different sources (proposed by WaM-DaM), we choose 10 water sources (rivers, lakes and reservoirs) and link their water height with precipitation and drought data available for their areas. We believe this exemplifies the utility of combining data from different sources in order to take water management related decisions.

The first objective of this chapter is to combine 3 different water management data sources (water level, precipitation and drought severity). We plot the data for our selected water sources and compute the minimum, maximum and average for each dataset.

Our second objective is to observe if there is any correlation between the water height values and precipitation/drought data respectively. In order to do this, we calculate the Pearson coefficients of correlation for measurements of the 3 parameters taken on dates in close proximity.

The chapter has the following structure. Section 12.2 describes related work in the water management field. In Section 12.3, we present the architecture used to ingest, process and visualize the data that we used in our model. We also concisely describe the structure of the datasets we used. In Section 12.4, we provide two possible real-life utilizations of a water resource management data model, with examples from [299]. Section 12.5 describes our datasets and implementation with more details. We also provide examples of plots for our datasets. In Section 12.6, we show the results of our experiments: processing time for each step in our model and values we consider important for our datasets (average, minimum, maximum, Pearson correlation coefficients). Finally we present our conclusions and provide possible ways our model could be extended to the future.

12.2   Related work

12.2.1   Data models

The Water Management Data Model (WaMDaM) [8] organizes and integrates water management data coming from different sources and in different formats. By eliminating the heterogeneity of the many data sources involved, WaM-DaM can be used to query the data from different sources and datasets more efficiently. Figure 12.1 shows the WaM-DaM architecture. Table 12.1 lists the features incorporated in WaMDaM and shows other models that provide some of those features.

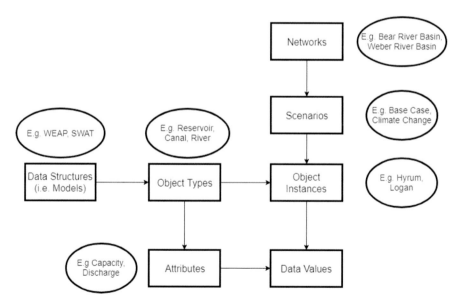

Figure 12.1: WaM-DaM architecture.

Table 12.1: WaM-DaM features.

Model	Flexible & extensible	Networks	Scenarios	Query with conditions	Dynamic controlled vocabulary	Explicit metadata	Multiple formats	Open source
WaDE				✓				✓
ODM-CUAHSI	✓			✓	✓	✓		✓
WEAP		✓	✓				✓	
GoldSim		✓	✓				✓	
WISKI Kisters							✓	
RiverWare								
GSSHA		✓						
SWMM		✓	✓				✓	✓
HEC-DSS		✓	✓				✓	
ArcSWAT		✓	✓				✓	
Arc Hydro	✓	✓		✓	✓	✓	✓	
CALVIN		✓	✓					✓
TOPNET			✓				✓	✓
AdHydro		✓					✓	✓
HydroPlatform	✓	✓					✓	✓

12.2.2   Datasets for water management

In this section, we describe 8 potential datasets that can be used in a water management system. Most of these datasets are in text format, each with their own structure, timespan, and location. Here, we provide a brief description and comparison of the datasets, while in Section 3—"Proposed architecture" we elaborate the analysis for the datasets we used in our model. Table 12.2 shows the datasets we investigated and their features.

For dataset number one, water levels of rivers and lakes are provided by the Equipe Géodésie, Océanographie & Hydrologie Spatiales (GOHS), part of Laboratoire d'Etudes en Geodésie et Océanographie Spatiales (LEGOS). The altimetry data for rivers comes from the Topex/Poseidon mission, while data from the ERS-1 and ERS-2, Envisat and Jason-1 missions was also used for the level of lakes. The altimetry data can be accessed after completing a form, while the gravimetry data for the river basins is accessible freely. Figure 12.2 shows the

Table 12.2: Dataset analysis.

Dataset number	Institution	Description	Period	Updated
1	LEGOS, GOHS	Water level of rivers and lakes[1], Altimetry data, open source for gravimetry (rivers), Text.	2002-2009	NO
2	USDA, FAS, IPAD	Lake height variations[2], Open source, Text.	1992-present	Every 10 days
3	DGFI, TUM	Water levels, Surface Area, Volume [316]. On request per each individual river/lake, Text.	1992-present	Every 10 days
4	USGS	Dynamic Surface Water Extent [186] [185], Open source, Images.	1982-present	Every 7 days
5	CRU UEA	Self-calibrating Palmer Drought Severity Index (scPDSI [363] [53], Open source, Text/NetCDF.	1901-2018	NO
6	CRU UEA	Global Land Precipitation [156], Open source, Text/NetCDF	1901-2018	NO
7	FAO	Water Resources, Use, Wastewater, Irrigation and Drainage [205], Open source, CSV.	1958-2017	YES
8	ANM	Precipitation [14], On request, Text.	1895-present	YES

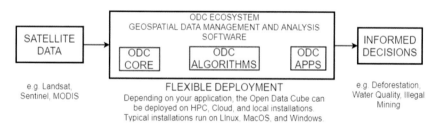

Figure 12.2: ODC architecture.

variation of Danube Basin water level, made available by the owners of this dataset.

For dataset number two, lake height variations are provided by the International Product Assessment Division (IPAD), part of the Foreign Agricultural Service (FAS) of the United States Department of Agriculture (USDA). This dataset uses satellite altimetry data from the Topex/Poseidon, Jason-1, Jason-2/OSTM, Jason-3, ERS-1 and ERS-2, Envisat, SARAL and Sentinel-3A missions. Data is uploaded periodically, usually for a 10-day window, starting in 1992.

For dataset number three, water level, surface area and volume from the Database for Hydrological Time Series of Inland Waters (DAHITI) are provided by the Deutsches Geodätisches Forschungsinstitut (DGFI) of the Technische Universität München (TUM). For approximating the water levels of rivers and lakes, satellite altimetry data from the Topex/Poseidon, Jason-1, Jason-2/OSTM, Jason-3, ERS-1 and ERS-2, Cryosat-2, IceSAT, SARAL, Sentinel-3A and Sentinel-3B missions is used. To access the data, the user must make an individual request for each river/lake.

For dataset number four, dynamic surface water extent is provided by United States Geological Survey (USGS), US Department of the Interior. The data can be accessed through the Earth Explorer interface provided by USGS, by selecting Datasets -¿ Landsat -¿ Landsat Collection 1 Level 3 -¿ Dynamic Surface Water Extent in the left side menu. The user can select the areas of interest from the map or from the menu on the left (data is only available for the US). The data can be downloaded by anyone with a registered account and is updated every 7 days, starting in 1998. The data comes in image format, with a product guide that explains what each pixel value represents.

For dataset number five, self-calibrating Palmer Drought Severity Index (scPDSI) data is provided by the Climatic Research Unit (CRU) from the University of East Anglia (UEA). This data set contains global scPDSI data of every month from 1901 to 2018 with a resolution of 0.5 degrees latitude and longitude and is described in the README. For the scPDSI, lower values show a more severe drought, with a value of -4 or below describing extreme drought conditions. Further information on the scPDSI can be found here [374] The data is made available under the Open Database License. Any rights in individual contents of the database are licensed under the Database Contents License.

For dataset number six, Global Land Precipitation data is also provided by the CRU unit of the University of East Anglia. There are two global datasets for the period between 1901 and 2018, with a resolution of 0.5 degrees latitude and longitude. At the time of this writing, the most recent version of this dataset is 4.03. The data is made available under the Open Database License. Any rights in individual contents of the database are licensed under the Database Contents License24.

For dataset number seven, water resources, water uses, wastewater, irrigation and drainage data are provided by the Food and Agriculture Organization (FAO) of the United States. The user can query a MySQL database or download data in CSV format. For each country, the user can also see a sheet with summary information in pdf format. The data is updated periodically, depending on the information available for each region.

For dataset number eight, precipitation data is provided by the Romanian National Weather Administration (ANM). The data from weather stations is available for download after an e-mail request. The update interval varies with each weather station.

12.2.3 Existing technologies

The Open Data Cube (ODC) is an open source framework used for the ingestion and analysis of satellite data (images). It consists of Python libraries and stores data in a PostgreSQL database. The ODC architecture can be found in Figure 12.2. The steps required for visualizing and analyzing data using the ODC are :

- ◼ Creating a product definition, which includes metadata for entire datasets (such as the dataset name, description and any measurements associated with the dataset). For each measurement (band), the datatype, a default value corresponding to no data and the unit of measurement must be provided.

- ◼ Indexing data, using a metadata file (in the .yaml format) for each scene. The metadata file contains information such as the coordinates of the polygon contained in the scene and the location of the images (in the .tif format) corresponding to each measurement. The ODC will match the scene to the product definition created in the previous step. After this step, a new entry will be created in the database for the added scene. The user can also index data available on AWS S3 (without having to download the scenes).

- ◼ Ingesting data (optional), which transforms the data into a more accessible format, that can be used for resampling or reprojection. In this step, the user can provide attributes such as ingestion bounds, and resolution.

CKAN is an open source tool used by organizations (companies, governments) to publish their data. The datasets published by these organizations can be viewed by users (the organization's staff, researchers). Each dataset published within CKAN has associated metadata and the data itself, which is held in one or more units called "resources". The resources can have any format or can be just a link to some data located elsewhere. Usually, datasets can be viewed without having to login, each user within an organization having a set of permissions over datasets (delete, edit or add other datasets). Figure 12.3 shows the CKAN architecture.

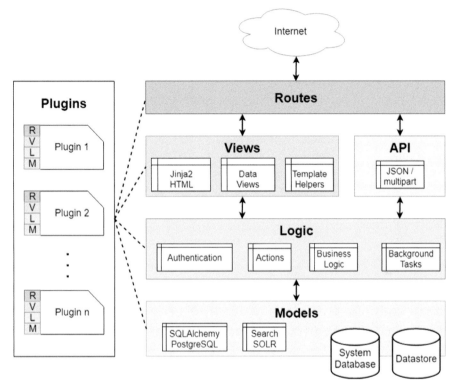

Figure 12.3: CKAN architecture.

12.3 Proposed architecture

12.3.1 *High level architecture*

Our architecture, viewed at the highest level, can be found in Fig. 12.4. The first part, "Datasets for water management", was described previously. In our implementation, we use 3 of the datasets from Table 12.2: water levels of rivers and lakes, scPDSI (Self-Calibrating Palmer Drought Severity Index) and precipitation data – lines 3, 5 and 6 from the table respectively. Of course, any other dataset can be used in a similar solution, with the appropriate adaptations of the code used for ingestion, processing and visualization.

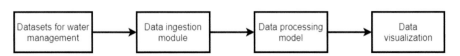

Figure 12.4: High level architecture.

Since the data source we selected for water levels was updated periodically (usually every 7 or 10 days, but the period varies with each water source), we decided to use Apache NIFI for a daily GET HTTP request and save the data for each river/lake we analyzed. Apache NIFI is a tool used to build and manage dataflows between multiple components of a system in a fault-tolerant manner. We selected 10 water sources (rivers, lakes and reservoirs) for our implementation, but additional sources can be easily added.

The complete architecture is shown in Fig. 12.5. Each part of the high-level architecture (ingestion, processing and visualization) is further described in its subsection.

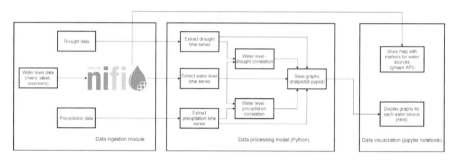

Figure 12.5: High level architecture.

12.3.2 *Data ingestion module*

Our Apache NIFI flow consists of an Invoke HTTP Processor for each of the 10 water sources (rivers, lakes and reservoirs) we have analyzed. Each of these Processors sends a daily GET request for its associated water source and forwards the response as a Flow File towards an Execute Stream Command Processor ("Flow Files" and "Flow File Processors" are some of the core concepts of Apache NIFI). To add another data source, the developer only has to add another Invoke HTTP Processor and configure it to request data for the desired river, lake or reservoir.

The Execute Stream Command Processor saves the data in a text file with the name and type of the water source (for example "Danube River.txt"). When testing the implementation, we configured the Invoke HTTP Processors to send GET requests every minute, but for the final solution we consider a daily interval to be enough for ingestion. In other use cases (where other datasets may be used), the developers can easily configure this interval at the desired value from the NIFI interface (if needed, different intervals can be chosen for each water source).

For the precipitation and drought data, we downloaded the data for the 1901-2018 period. Each dataset was a text file with monthly values for the precipitation or scPDSI of a global grid, with a resolution of 0.5 degrees latitude and longitude. For each month, the dataset had 360 lines of data, one for each latitude line in the grid (from the areas with the center at 89.75°N latitude to 89.75°S latitude). Each line consists of 720 values (from 179.75°W to 179.75°E), with no data values encoded as -9999 for scPDSI and -999 for precipitation. Each consecutive month is appended to the file, from January 1901 to December 2018.

To make the data for our analyzed water sources easier to access, we broke the precipitation and scPDSI files (which had a size of about 3GB each) to 360 files, for each latitude line in the 0.5° resolution grid. This way, we only have to process a few of those files (of about 10MB each), corresponding to the areas where our selected water sources are located.

12.3.3 Data processing model

Our processing model is implemented in Python, using Jupyter Notebooks to visualize a map generated with the gmaps API. The first step in our model is processing the water height information for the 10 water sources we selected. We read the data from each file ingested using NIFI and extract some useful information (such as the water source name, type, location and average height) from the start of the file (first 10 lines) into a Python dictionary we call "metadata". This metadata is very important in the model, as we use it in every other step of the processing. The actual height data is also stored in a Python dictionary, with the timestamps of the data as keys and the water heights as values. In some of the original files, the height data is available as height variation relative to the average we store in our metadata, in others the height is given as an absolute value and we compute the average.

Next, we process the precipitation and scPDSI information. The procedure is similar for both datasets, as the input files are very similar. For each water source, we find the line in the grid described in the previous section in which our lake, river or reservoir is located. We use the data only for that particular line, reading the corresponding file we created in the ingestion phase. We create a monthly time series as a list of values, from January 1901 to December 2018 and a yearly time series as a list of values for the 1901-2018 period. For the precipitation dataset, the yearly values are estimated as the monthly average for the months when data is available multiplied by 12. For the drought data, the yearly values are the average values for the monthly data available for that year. In both cases, if there is no available data for an entire year, that year's value is equal to the no data value (-999 for precipitation and -9999 for drought).

The next step is to correlate the datasets. We correlate the height information with both the precipitation and drought data by calculating the Pearson correlation coefficient. In order to do this, we keep only a subset of the height dataset, for the months we have available precipitation or drought data. We also use only a subset of the precipitation and drought timeseries, because for the water level we only have recent measurements (each water source has a different data for the first measurement, but all start after 1992). We construct two pairs of Python lists: (height, precipitation) and (height, drought). In each pair, the lists have the same length. The values on corresponding positions in the lists have the same timestamp. This way, we can easily plot the two lists and visually compare them, as they have the same time axis. In each pair, the lists contain values only for the dates when we have both height and precipitation (or drought) data available. Having data paired by timestamp also made it easy to compute the Pearson coefficients, using the pearsonr function from the scipy Python package.

After taking the above steps for each of our 10 water sources, the correlation coefficients indicated almost no correlation: except for the Danube River, which had a 0.58 value for the Pearson coefficient for the height-precipitation correlation, the second highest value was approximately 0.25, for the height-drought correlation of the Rybinsk Reservoir. Although disappointing at first, at a second thought this result was somehow expected. Because we investigated major water sources, their height is not significantly influenced by a period of heavy rain or severe drought (a relatively important river or lake will not disappear after a year with little rain).

The last processing step is saving graphs for each water source with the height, precipitation and drought data, as well as the graphs of the time series used for correlations: height-precipitation and height-drought. We also generate a html page where all these graphs are shown.

12.3.4   Data visualization

For visualizing the data, we use Jupyter Notebooks and the gmaps API. For each water source, we extract the location (latitude and longitude values) from the metadata described in the previous section. In the notebook, we display a map with markers in the locations of our rivers, lakes and reservoirs. The user can click these markers to view the water source's summary information: name, type and minimum/maximum/average height for the entire period available in that location. A link is also available, which opens the html page, which was previously generated, showing all the graphs in a new browser tab.

12.4   Examples of real-life utilizations

12.4.1   Flooding alerts

Water level monitoring is critical for reducing the risks and effects of floods. By combining satellite-collected data with other data sources, it is now possible

Figure 12.6: Map in Jupyter notebooks.

to better monitor parameters of water sources, such as height, extent or quality. Floods can occur not only after a period of heavy rain, but also during spring, if the temperature increases rapidly and snow melts at a high rate, resulting in high quantities of water being brought into rivers. With current developing climate changes, these kinds of events are occurring more often. Thus, monitoring the parameters of water systems on a nearly daily basis (which became feasible with the collection of satellite images) becomes more and more important for understanding and reducing the impact of floods.

A data model similar to the one presented in this thesis can help monitor and perhaps reduce the risks of floods by alerting authorities when such a risk is imminent. First, by observing the level of water sources over an extended period provides a better understanding of the water source's height variation pattern throughout the year. After observing an abnormal level variation (for example, the height exceeds a value that was observed to lead to floods in the past), an alert can be generated.

Next, authorities can evaluate the risk by taking into account other data sources existing in the model. For example, precipitation data for the period before the alert signal was generated, along with the correlation of height-precipitation for our water source in the past can provide a better understanding of the rate at which the water level will increase in the near future. Furthermore, the weather forecast for the next few days might determine the authorities to prepare to intervene in populated areas around the observed water source.

An example of flood monitoring using satellite data can be found in the book "The ever growing use of Copernicus across Europe's regions: a selection of 99 user stories by local and regional authorities" [299]. This article [279]describes the confrontation with the flood in Ireland in 2015/2016. Because almost no hydrometric data was documented, observations provided by the Sentinel-1 program was critical. As stated in the article, "the Synthetic Aperture Radar (SAR) capability of the Sentinel-1 constellation is particularly valuable because of its ability to detect differences in land cover and provide an all-weather, day-and-night supply of imagery". Collecting data every few days is enough to monitor a groundwater flood, which takes a longer time to develop (weeks or even months, in contrast with flash floods, which evolve in hours).

12.4.2   Helping water resource managers

A model that connects many data sources can be particularly useful for water resource managers, who are interested not only in the water levels of rivers and lakes, but also in the quality of water and planning of water use. In [306] describing the aid provided by satellite data (Sentinel-2 and Landsat 8) for water resource management. Mulargia is an important water system in Sardinia, and the local water manager must provide information to the drinking water supplier about the pollution level and amount of water available.

Satellite information completes the ground systems in the computation of indicators regarding water quality. According to the article, "these indicators include: Chlorophyll-a, turbidity and harmful algae blooms; water surface temperature and evaporation; floating materials (for example, oil or scum)" [306]. By combining satellite imagery with traditional data collection tools, water managers can now make more accurate predictions regarding threats to the quantity and quality of the water sources they observe. Furthermore, by measuring more features of water sources, the consumption can be optimized for local communities and industries.

12.5   Implementation details

12.5.1   Implementation of the data ingestion module

As we described in Section 12.3, our ingestion module is in charge of collecting 3 types of data:

First is water levels of the rivers, lakes and reservoirs we selected: Danube, Po and Prypjat rivers; Balaton, Ilmen, Ladoga, Onega and Peipus lakes; Kremenchuk and Rybinsk reservoirs. Since these datasets are updated periodically, we decided to use Apache NIFI for the ingestion of data. Our NIFI Flow consists of 10 Invoke HTTP Processors (one for each water source) and an Execute Stream Command Processor, which receives the HTTP responses and saves the files. Each NIFI Processor is configured with the address of the dataset it will get, username and password for the website and time between consecutive GET requests. For testing, we used a one-minute timer, but we recommend a daily request for each desired water source.

The first 10 lines of each file contain useful information about the water source: name, type (river, lake or reservoir), latitude, longitude and sometimes average height. The data lines have two fields that we use in our model: the first field is the date, while the second field represents the height. For the datasets that contain the average height as a metadata field, the heights are given relative to the average. For the others, we compute the average and add it to the metadata dictionary.

The developer can add a new water source by simply modifying the NIFI Flow and adding a new Invoke HTTP Processor, configured to get data for the desired water source. After linking the new processor to the Execute Stream Command Processor and starting the flow, the new processor will automatically save the HTTP responses containing the desired data with the periodicity configured.

The file of global precipitation data, with a resolution of 0.5 degrees latitude and longitude for the 1901-2018 period, consists of 360 lines for each month (from January 1901 to December 2018), each line describing the precipitation values of a latitude line in the 0.5 resolution grid. The first line for each month contains values for the line with the center at $89.75N$ latitude, while on the last line we find values for the line at $89.75S$ latitude. The next month is then appended to the file and so on. On each line, the first value corresponds to the area with the center at $179.75W$ longitude and the value of $179.75E$ longitude as the last field.

Each value is measured in mm of precipitation multiplied by 10 (for example a value of 175 means a precipitation of $17.5mm$ in that area for that month), with values of -999 meaning no data is available (in the case of ocean areas, for example). The translation between measurement units is as follows: 1 mm of precipitation is equivalent to $1l/m^2$ and signifies a water height of 1 mm if the precipitation is equally distributed on an area made of an impermeable material.

We break down the original file into 360 files, one for each latitude line in the grid. This action is quite time consuming, but it is only done once after the download of the original file (with a size of 2.9GB). In the processing phase, we only use data for the lines where our selected water sources are located and optimize the processing time by working with much smaller files (of 8.2MB each), containing just a subset of the data.

The dataset of global scPDSI (Self-Calibrating Palmer Drought Severity Index) is very similar to the precipitation data: it has a resolution of 0.5 latitude and longitude and covers the 1901-2018 period. The file also has 360 lines for each month, with 760 values on each line and the next month appended to the file. The lines and values within each line correspond to the same areas as their precipitation counterparts. The no data value for this dataset is -9999 (instead of -999), and the values represent the monthly scPDSI, with a value of 4 or more meaning extreme wet conditions and a value of -4 or less meaning extreme drought conditions. Similar to the precipitation data, this file (3.7GB) is broken down into 360 files of 10.2MB each (one for each latitude line).

The main problem with the PDSI was that it was unable to construct comparisons, as its meaning was location-dependent (the same value of the index could mean different drought severities, depending on other climatic conditions in the area) [374]. The scPDSI is calculated using data about temperature, precipitation and soil Available Water Content (AWC).

12.5.2 Implementation of the processing model

For the water height data, we use the first 10 lines of each file to generate a "metadata" for each water source. The important pieces of information we use from the metadata are: water source name, type, latitude, longitude and, for some water sources, the average height. We represent the metadata as a Python dictionary, where we add the average height we computed (for datasets with absolute heights provided) and the path to the html page is displayed in the visualization phase (derived from the name and type of the water source).

Next, we create another dictionary for the actual height information. On each line of data, the first field is the date, while the second field represents the height relative to the average we stored in the metadata. We add this difference to the average of each day for which we have data and store the results as values in the heights dictionary. The keys of the dictionary are datetime objects representing the first field of each line.

For each data source, we also compute the last year's height and average height. We keep from the heights dictionary only those (key, value) pairs for a year's period ending on the current day.

We include the height variations and the last year's height variations in a graph for each water source. On each graph, we display horizontal lines for the average, minimum and maximum height presented on that graph. Figure 12.7 shows the

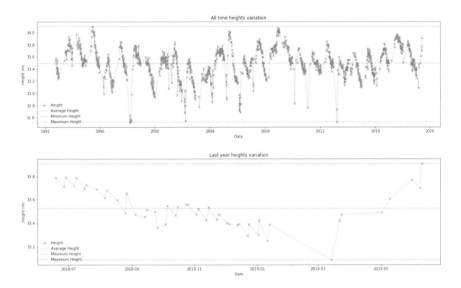

Figure 12.7: Height variations for the Onega Lake.

height graphs for the Onega Lake, with the data that was available on 15.06.2019 (data for the period 01.10.1992-11.06.2019).

The precipitation and drought data are processed in a similar manner, as the datasets are similar. For each water source, we read the file for the appropriate latitude line (the contents of the files are described in the previous section) and use the values for our longitude. We find the coordinates of the water source from the metadata we described earlier.

We construct 4 Python lists: monthly and yearly time series, with datetime objects for each month/year in the start year-end year interval (in our case, 1901-2018) and monthly and yearly precipitation/scPDSI data. The first value of the monthly precipitation list is the value for the first month (or -999, if no data is available) and so on, until the last month. Due to the structure of the datasets we use, it is guaranteed that the pairs of lists (monthly time series, monthly values) and (yearly time series, yearly values) will have the same length respectively. By representing the data this way, we do not add any complexity to our solution: if a developer wants to add a functionality to the model and needs to access a value for a certain month, the position in the list is easily computed as (year-start year)*12+month -start month. With this representation, it becomes easier to plot the data: we use the time series on the x axis and the values on the y axis.

The yearly values for the two datasets are computed differently. For the precipitation data, the total yearly precipitation is estimated as monthly average*12, where monthly average is the average of the months for which we have data

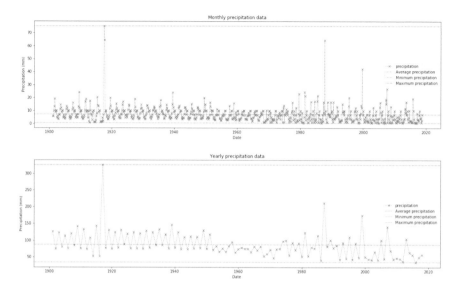

Figure 12.8: Onega Lake precipitation.

available. For the drought data, we compute the average scPDSI as the average of the monthly values we have data available. In both cases, if there is no available data for the whole year, we insert a no data value for that year (-999 for precipitation and -9999 for scPDSI).

Next, we save graphs for precipitation and drought data. In order to do this, we extract from the monthly and yearly values only those for which represent actual data. We iterate through the list of monthly values and keep only those values which are different from the no data value. These values are on the y axis of our plot. On the x axis, we select from the monthly time series the dates for which we have data (we find those dates on the same indices in the list as the values we kept). We also plot the minimum, maximum and average precipitation/scPDSI as horizontal lines. A similar procedure is followed for the yearly data: we select values for the years we have at least one month of valid data and display a graph with the estimated annual precipitation and the average annual scPDSI, with the same horizontal lines for average, minimum and maximum. Figure 12.8 shows the precipitation graphs for the area surrounding the Onega Lake, while Fig. 12.9 shows the variation of the scPDSI for the same area.

We also correlate the height data with both the precipitation and drought datasets. In order to do this, we extract the values corresponding to the timestamps for which we have both a height measurement and a precipitation value that is valid and we generate 3 lists. The first list contains datetime objects representing the common time axis for our correlation. This list is a subset of the keys from the heights dictionary (described at the beginning of this chapter) with dates for

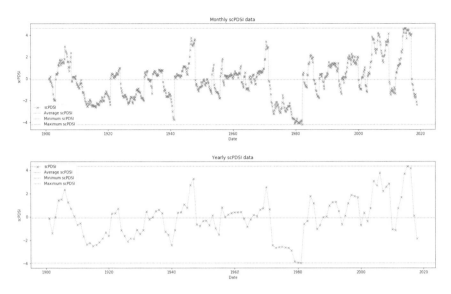

Figure 12.9: Onega Lake scPDSI.

which we also have valid precipitation data. The second and third lists contain the water heights and precipitation values respectively. The list of heights is simply the values (from the heights dictionary) of the keys we kept in the first list. It is possible that more than one height measurement will be available for the same month. In this case, in the precipitation list we will have the value for that month appearing more than once, to keep both the heights and precipitation lists with the same time axis. For the drought data, we follow the same procedure we described for precipitation. We plot the lists we wish to correlate with the same x (time) axis, and we compute the Pearson correlation coefficients. In Fig. 12.10 we have the heights and precipitation graps for the Danube River (with a Pearson coefficient of approximately 0.58) while in Fig. 12.11 we show the heights and scPDSI graphs for the Rybinsk Reservoir (with a correlation coefficient of approximately 0.25).

12.6   Experimental results

In this section, we analyze the performance of our implementation. The highest overhead was caused by the splitting of the precipitation (2 minutes and 21 seconds) and drought (2 minutes and 42 seconds) data files into separate files for each latitude line (details about this procedure were provided in Section 5.1). However, this operation only occurs once a month, as the precipitation and scPDSI (Self-Calibrating Palmer Drought Severity Index) data is gathered for a one-month period and is appended to the previous file. The data processing

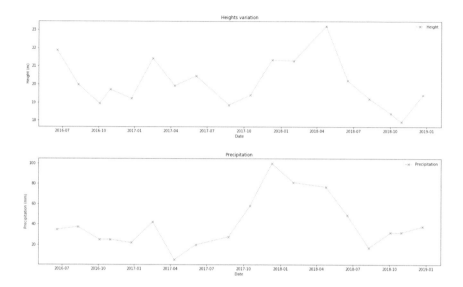

Figure 12.10: Onega Lake scPDSI.

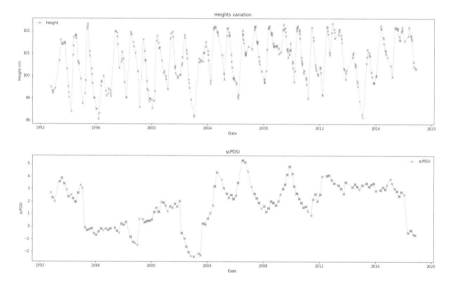

Figure 12.11: Onega Lake scPDSI.

(described in Section 5.2) and saving the graphs for visualization is done only once a day, as new height data is ingested daily (through NIFI). The processing and saving of the graphs for each water source takes approximately 2 seconds. In Table 12.3 we list the times for the processing of the water sources (including

Table 12.3: Total and average processing time for each phase.

	Average time per water source (seconds)	Total time for 10 water sources (seconds)
Height processing	0.41	4.11
Precipittion processing	0.51	5.10
Drought processing	0.49	4.91
Height-precipitation correlation	0.35	3.48
Height-drought correlation	0.35	3.49
Total	2.11	21.09

saving the graphs). The values in the table represent the average of 10 experiments. For each processing phase, we list both the total time (for all the 10 water sources we analyzed) and the average time per water source.

After all the processing is done, the data can be analyzed (by water resource managers, researchers) for a whole day, until new information is brought into the system. Because the implementation of our model (or a similar one) would most likely be used by local authorities or water resource managers (who usually handle a single river basin or reservoir), the processing can be done when data is not being analyzed (for example, at night). Alternatively, if we use the model to take into account many water sources, their corresponding NIFI processors can be configured to ingest data at different times and we can process the data immediately after each water source height variation has been updated. This way, people who want to visualize the data will never have to wait for all the water sources to be processed at once. For the NIFI flow, a task takes approximately 1 second for each of the 11 processors (10 Invoke HTTP Processors and the Execute Stream Command Processor).

12.6.1 Analysis of experimental results

As we described in Section 5.2, for each river, lake and reservoir we selected for our experiments, we computed the minimum, maximum and average for the parameters we took into account (water level, precipitation and scPDSI). We also tried to offer the user the possibility to correlate water height with both the precipitation and the drought data. This correlation can be done either empirically (by visualizing two graphs with the same time axis), or by observing the Pearson correlation coefficients we computed for the two correlations for each water source. We list all the results for water heights (for all available measurements and for the last year) in Table 12.4, where for the Po River we have no measurements for the last year. In Table 12.5 and Table 12.6 we show the monthly and yearly results for precipitation and drought data processing respectively, as well as the Pearson correlation coefficients. As previously discussed in Section

Table 12.4: Results for water height processing.

Water source	First measure-ment	Last mea-surement	Average height (m)	Min. height (m)	Max. height (m)	2019 average height (m)	2019 min height (m)	2019 max height (m)
Balaton Lake	14.08.08	04.06.19	104.721	103.908	105.460	104.771	103.972	105.432
Danube River	18.06.16	10.05.19	20.154	17.912	23.216	19.598	17.912	21.430
Ilmen Lake	06.07.02	05.04.19	18.586	15.743	21.620	17.796	17.070	19.285
Kremenchuk Reservoir	04.10.92	02.06.19	80.344	76.753	81.548	80.606	78.522	81.529
Ladoga Lake	25.09.92	08.06.19	4.801	3.029	6.095	5.144	4.602	5.912
Onega Lake	01.10.92	11.06.19	33.492	32.520	34.100	33.525	33.086	33.906
Peipus Lake	14.02.02	01.06.19	30.041	28.759	31.294	29.762	29.442	30.173
Po River	12.08.02	01.06.16	19.719	16.908	24.861	-	-	-
Prypjat River	23.07.08	02.06.19	116.486	114.381	118.641	114.786	114.897	117.199
Rybinsk Reservoir	25.09.92	03.06.19	100.751	97.884	102.424	100.943	99.518	101.996

Table 12.5: Results for precipitation processing and correlation with height variation.

Water source	Avg monthly precipita-tion (mm)	Min monthly precipita-tion (mm)	Max monthly precipita-tion (mm)	Avg yearly precipita-tion (mm)	Min yearly pre-cipitation (mm)	Max yearly pre-cipitation (mm)	Pearson coeffi-cient
Balaton Lake	103.9	7.0	354.8	1246.7	610.8	2042.8	0.04
Danube River	50.4	0.7	174.4	604.5	376.2	893.8	0.58
Ilmen Lake	26.2	0.0	224.4	313.9	38.4	817.0	-0.20
Kremenchuk Reservoir	52.7	0.4	148.3	623.3	247.2	1104.8	-0.23
Ladoga Lake	0.9	0.0	8.8	10.6	1.4	31.4	0.05
Onega Lake	7.3	0.4	75.4	87.3	34.0	325.2	0.00
Peipus Lake	19.5	0.0	192.2	233.7	43.2	676.2	-0.09
Po River	70.5	0.0	270.6	846.2	383.2	1458.8	0.07
Prypjat River	62.8	0.1	332.7	753.1	338.4	1236.0	0.1
Rybinsk Reservoir	4.7	0.0	26.3	56.5	9.4	127.6	0.04

3.3, we found no correlation for most water sources. We mention that the period for precipitation and drought data is January 1901–December 2018 for all water sources.

12.7 Conclusions and future work

In this thesis, we show the utility of the Water Management Data Model (WaM-DaM) by integrating 3 heterogenous data sources: water height, precipitation and

Table 12.6: Results for drought processing and correlation with height variation.

Water source	Avg monthly scPDSI	Min monthly scPDSI	Max monthly scPDSI	Avg yearly scPDSI	Min yearly scPDSI	Max yearly scPDSI	Pearson coefficient
Balaton Lake	0.54	-3.90	4.63	0.54	-3.31	4.26	-0.16
Danube River	0.21	-5.06	5.60	0.21	-4.55	5.47	0.07
Ilmen Lake	0.70	-4.38	5.30	0.70	-3.93	4.83	-0.06
Kremenchuk Reservoir	-0.97	-5.30	4.39	-0.97	-4.91	3.98	-0.01
Ladoga Lake	0.06	-4.29	6.17	0.06	-4.01	5.16	0.09
Onega Lake	-0.10	-4.23	4.62	-0.10	-3.96	4.33	-0.09
Peipus Lake	-0.18	-4.88	5.83	-0.18	-3.19	2.85	0.04
Po River	-0.78	-5.56	4.64	-0.78	-5.19	3.71	0.09
Prypjat River	0.85	-4.81	5.68	0.85	-4.67	5.08	-0.18
Rybinsk Reservoir	0.99	-5.88	5.26	0.99	-5.45	4.25	0.25

drought data. For each of the 10 water sources (rivers, lakes and reservoirs) we selected, we provide the user with information and graphs about the parameters we analyzed. We also present two examples of real-life utilization, described in detail in Section 4. One refers to the possibility of predicting an imminent flood by analyzing historical data for a local water source. The other shows the support towards resource planification given to water resource managers and authorities by a combination of satellite-collected and traditionally-gathered data.

Another objective is to observe possible correlations between water level and precipitation/drought data respectively. As we stated in Section 3 and Section 5, no correlation was observed, except for the Danube River height-precipitation correlation. This result is most likely caused by the fact that we analyzed major water sources, whose water heights don't drop dramatically after a period of severe drought.

In the future, we can integrate our model with satellite-collected data, using the Open Data Cube framework. An organization using our model can also use CKAN to list metadata describing the datasets used in the model.

References

[1] Fire safety risk assessment: small and medium places of assembly, Jun 2006.

[2] 5 reasons to use workplace desk, room and office occupancy monitoring, May 2020.

[3] S. Ren A. S. M. Hasan Mahmud. Online Resource Management for Data Center with Energy Capping. In *8th International Workshop on Feedback Computing (Feedback Computing 13)*. USENIX Association, 2013.

[4] AAA. Automatic emergency braking with pedestrian detection. Technical report, American Automobile Association (AAA), Florida, United States, October 2019.

[5] Daniel J. Abadi, Peter A. Boncz, and Stavros Harizopoulos. Column-oriented database systems. *Proceedings of the VLDB Endowment*, 2(2):1664–1665, 2009.

[6] Arndt Bonitz Abdelkader Magdy Shaaban, Christoph Schmittner. The design of a divide-and-conquer security framework for autonomous vehicles. In *The Eighth International Conference on Advances in Vehicular Systems, Technologies and Applications*, pages 49–102, 2019.

[7] Assad Abu-Jasser and Husam Shaheen. Voltage-control based on fuzzy adaptive particle swarm optimization technique. *Emirates Journal for Engineering Research*, 16(1), 2011.

[8] David Rosenberg Adel Abdallah. Wam-dam: A data model to organize and synthesize water management data. In *Environmental Modelling and Software (iEMSs)*, 2014.

[9] NHTSA National Highway Traffic Safety Administration. Assessment of safety standards for automotive electronic control systems. Technical report, United States. National Highway Traffic Safety Administration, 2016.

[10] Wim Aertsen, Vincent Kint, Klaus Von Wilpert, Dietmar Zirlewagen, Bart Muys, and Jos Van Orshoven. Comparison of location-based, attribute-based and hybrid regionalization techniques for mapping forest site productivity. *Forestry: An International Journal of Forest Research*, 85(4):539–550, 2012.

[11] ETP4HPC Agenda. European technology platform for high performance computing. *Strategic research agenda achieving HPC leadership in Europe, www.etp4hpc.eu.*

[12] AgingInPlace. Technology in our life today and how it has changed. https://www.aginginplace.org/technology-in-our-life-today-and-how-it-has-changed/, 2019. (accessed January 28, 2020).

[13] G Ahalya and Hari Mohan Pandey. Data clustering approaches survey and analysis. In *2015 International Conference on Futuristic Trends on Computational Analysis and Knowledge Management (ABLAZE)*, pages 532–537, 2015.

[14] Jamal Ahmadaali, Gholam-Abbas Barani, Kourosh Qaderi, and Behzad Hessari. Analysis of the effects of water management strategies and climate change on the environmental and agricultural sustainability of urmia lake basin, iran. *Water*, 10(2):160, 2018.

[15] Neethu Akkarapatty and Nisha S Raj. A machine learning approach for classification of sentence polarity. In *2016 3rd International Conference on Signal Processing and Integrated Networks (SPIN)*, pages 316–321, 2016.

[16] B. Aksanli, J. Venkatesh, L. Zhang, and T. Rosing. Utilizing green energy prediction to schedule mixed batch and service jobs in data centers. In *Proceedings of the 4th Workshop on Power-Aware Computing and Systems*, page 1–5. ACM, 2012.

[17] Imperial College London AMCG. Fluidity manual v4.1.12. 4 2015.

[18] Samaneh Aminikhanghahi and Diane J. Cook. A survey of methods for time series change point detection. *Knowledge and Information Systems*, 51(2):339–367, 2016.

[19] The VERGE Andrew J. Hawkins. Cars with high-tech safety systems are still really bad at not running people over. https://www.theverge.com/2019/10/4/20898773/aaa-study-automatic-emergency-braking-pedestrian-detection, 2019. (accessed January 23, 2020).

[20] G. Andrienko, N. Andrienko, P. Jankowski, D. Keim, M.-J. Kraak, A. MacEachren, and S. Wrobel. Geovisual analytics for spatial decision support: Setting the research agenda. *International Journal of Geographical Information Science*, 21(8):839–857, sep 2007.

[21] G. Andrienko, N. Andrienko, and R. Weibel. Geographic data science. *IEEE Computer Graphics and Applications*, 37(5):15–17, 2017.

[22] G. Andrienko, N. Andrienko, and R. Weibel. Geographic data science. *IEEE Computer Graphics and Applications*, 37(5):15–17, 2017.

[23] Ionut Anghel, Massimo Bertoncini, Tudor Cioara, Marco Cupelli, Vasiliki Georgiadou, Pooyan Jahangiri, Antonello Monti, Seán Murphy, Anthony Schoofs, and Terpsi Velivassaki. Geyser: Enabling green data centres in smart cities. In *International Workshop on Energy Efficient Data Centers*, pages 71–86. Springer, 2014.

[24] R Arcucci, L Carracciuolo, and R Toumi. Toward a preconditioned scalable 3dvar for assimilating sea surface temperature collected into the caspian sea. *JOURNAL OF NUMERICAL ANALYSIS, INDUSTRIAL AND APPLIED MATHEMATICS*, 12(1-2):9–28, 2018.

[25] Rossella Arcucci, Luisa D'Amore, Jenny Pistoia, Ralf Toumi, and Almerico Murli. On the variational data assimilation problem solving and sensitivity analysis. *Journal of Computational Physics*, 335:311–326, 2017.

[26] Rossella Arcucci, Luisa D'Amore, Luisa Carracciuolo, Giuseppe Scotti, and Giuliano Laccetti. A decomposition of the tikhonov regularization functional oriented to exploit hybrid multilevel parallelism. *International Journal of Parallel Programming*, 45(5):1214–1235, 2017.

[27] Rossella Arcucci, Laetitia Mottet, Christopher Pain, and Yi-Ke Guo. Optimal reduced space for variational data assimilation. *Journal of Computational Physics*, 379:51–69, 2019.

[28] Rossella Arcucci, Laetitia Mottet, Cesar Quilodran Casas, Florian Guitton, Christopher Pain, and Yi-Ke Guo. Adaptive domain decomposition for effective data assimilation. In *Lecture Notes in Computer Science book series (EuroPAR 2019).* (in print), 2019.

[29] Rossella Arcucci, Christopher Pain, and Yi-Ke Guo. Effective variational data assimilation in air-pollution prediction. *Big Data Mining and Analytics*, 1(4):297–307, 2018.

[30] Rossella Arcucci, Cesar Quilodran Casas, Xiao Dunhui, Laetitia Mottet, Fangxin Fang, Pin Wu, Christopher Pain, and Yi-Ke Guo. A domain decomposition reduced order model with data assimilation (dd-roda). In *Advances in Parallel Computing*, volume 36, pages 189–198. DOI: 10.3233/APC200040, 2020.

[31] E. Aristodemou, T. Bentham, C. Pain, and A. Robins. A comparison of mesh-adaptive les with wind tunnel data for flow past buildings: mean flows and velocity fluctuations. *Atmospheric Environment journal*, 43:6238–6253, 2009.

[32] Elsa Aristodemou, Rossella Arcucci, Laetitia Mottet, Alan Robins, Christopher Pain, and Yi-Ke Guo. Enhancing cfd-les air pollution prediction accuracy using data assimilation. *Building and Environment*, 165:106383, 2019.

[33] Muhammad Arslan, Christophe Cruz, and Dominique Ginhac. Semantic enrichment of spatio-temporal trajectories for worker safety on construction sites. *Personal and Ubiquitous Computing*, 23(5-6):749–764, 2019.

[34] Muhammad Arslan, Christophe Cruz, and Dominique Ginhac. Semantic trajectory insights for worker safety in dynamic environments. *Automation in Construction*, 106:102854, 2019.

[35] Muhammad Arslan, Christophe Cruz, and Dominique Ginhac. Spatio-temporal dataset of building occupants. *Data in brief*, 27:104598, 2019.

[36] Muhammad Arslan, Christophe Cruz, and Dominique Ginhac. Understanding occupant behaviors in dynamic environments using obide framework. *Building and Environment*, 166:106412, 2019.

[37] Muhammad Arslan, Christophe Cruz, and Dominique Ginhac. Visualizing intrusions in dynamic building environments for worker safety. *Safety Science*, 120:428–446, 2019.

[38] Mark Asch, Marc Bocquet, and Maëlle Nodet. *Data assimilation: methods, algorithms, and applications*, volume 11. SIAM, 2016.

[39] Yudistira Asnar, Elda Paja, and John Mylopoulos. Modeling design patterns with description logics: A case study. In *International Conference on Advanced Information Systems Engineering*, pages 169–183. Springer, 2011.

[40] North Carolina Forestry Association. *Tree Measurement*. 2016.

[41] V Babovic, M Keijzer, and M Bundzel. From global to local modelling: a case study in error correction of deterministic models. In *Proceedings of fourth international conference on hydroinformatics*, 2000.

[42] Vladan Babovic, Rafael Cañizares, H René Jensen, and Anders Klinting. Neural networks as routine for error updating of numerical models. *Journal of Hydraulic Engineering*, 127(3):181–193, 2001.

[43] Vladan Babovic and David R Fuhrman. Data assimilation of local model error forecasts in a deterministic model. *International journal for numerical methods in fluids*, 39(10):887–918, 2002.

[44] Andrew Bakun and Kenneth Broad. Environmental 'loopholes' and fish population dynamics: comparative pattern recognition with focus on el nino effects in the pacific. *Fisheries Oceanography*, 12(4-5):458–473, 2003.

[45] Kasun Bandara, Christoph Bergmeir, and Slawek Smyl. Forecasting across time series databases using recurrent neural networks on groups of similar series: A clustering approach. *Expert Systems with Applications*, 140:112896, feb 2020.

[46] L.N. Barrett. Learning all optimal policies with multiple criteria. In *Proceedings of the 25 th International Conference on Machine Learning. Helsinky, Finland*. SIAM Press, 2008.

[47] J. D. Barroso. The tail at scale. *Communications of the ACM*, pages 74–80, 2013.

[48] P. Bellavista, S. Chessa, L. Foschini, L. Gioia, and M. Girolami. Human-enabled edge computing: Exploiting the crowd as a dynamic extension of mobile edge computing. *IEEE Communications Magazine*, 56(1):145–155, Jan 2018.

[49] R. Bellman. *Dynamic Programming*. Princeton, NJ,USA: Princeton University Press, 1957.

[50] Yann Bertacchini, Marisela Rodríguez-Salvador, and Wahida Souari. From territorial intelligence to compositive & sustainable system. case studies in mexico & in gafsa university. 2007.

[51] Ajay Gajanan Bhave, Declan Conway, Suraje Dessai, and David A Stainforth. Water resource planning under future climate and socioeconomic uncertainty in the cauvery river basin in karnataka, india. *Water resources research*, 54(2):708–728, 2018.

[52] A Birnie and T v Roermund. A multilayer vehicle security framework. *white paper, Date of release: May*, 2016.

[53] Jessica Blunden, Derek S Arndt, Gail Hartfield, Gesa A Weyhenmeyer, and Markus G Ziese. State of the climate in 2017. *Bulletin of The American Meteorological Society-(BAMS)*, 99(8):S1–S310, 2018.

[54] Andrea Bonci, Alessandro Carbonari, Alessandro Cucchiarelli, Leonardo Messi, Massimiliano Pirani, and Massimo Vaccarini. A cyber-physical system approach for building efficiency monitoring. *Automation in Construction*, 102:68–85, 2019.

[55] G. Booch. *Object-Oriented Analysis and Design with Applications.*

[56] Jeffry A. Borror. *Q For Mortals: A Tutorial In Q Programming.* CreateSpace, 2008.

[57] Brian Brazil. *Prometheus: Up & Running: Infrastructure and Application Performance Monitoring.* O'Reilly Media, Inc., 2018.

[58] Peter J. Brockwell and Richard A. Davis. *Time series: theory and methods.* Springer, 1991.

[59] Peter J. Brockwell and Richard A. Davis. *Introduction to Time Series and Forecasting.* Springer, 2016.

[60] G. Cacciatore, C. Fiandrino, D. Kliazovich, F. Granelli, and P. Bouvry. Cost analysis of smart lighting solutions for smart cities. In *IEEE International Conference on Communications (ICC)*, pages 1–6, May 2017.

[61] D.G. Cacuci, I. M. Navon, and M. Ionescu-Bujor. Computational methods for data evaluation and assimilation. *CRC Press*, 2013.

[62] A. Capponi, C. Fiandrino, B. Kantarci, L. Foschini, D. Kliazovich, and P. Bouvry. A survey on mobile crowdsensing systems: Challenges, solutions, and opportunities. *IEEE Communications Surveys Tutorials*, 21(3):2419–2465, Third Quarter 2019.

[63] Car 2 Car Communication Consortium. Protection Profile V2X Hardware Security Module. Protection profile, Car 2 Car Communication Consortium, 2019.

[64] Carrie Cox and Andrew Hart. How autonomous vehicles could relieve or worsen traffic congestion. Technical report, Here Technologies, 2017.

[65] CBInsights. The future of data centers. https://www.cbinsights.com/research/future-of-data-centers/, 2019. Accessed: 2019-07-10.

[66] Samarjit Chakraborty, Mohammad Abdullah Al Faruque, Wanli Chang, Dip Goswami, Marilyn Wolf, and Qi Zhu. Automotive cyber–physical systems: A tutorial introduction. *IEEE Design & Test*, 33(4):92–108, 2016.

[67] C.Hansen. Rank-deficient and discrete ill-posed problems, numerical aspects of linear inversion. *SIAM*, 1998.

[68] Himadri Chauhan, Vipin Kumar, Sumit Pundir, and Emmanuel S Pilli. A comparative study of classification techniques for intrusion detection. In *2013 International Symposium on Computational and Business Intelligence*, pages 40–43, 2013.

[69] H. Chen, R. H. L. Chiang, and V. C. Storey. Business intelligence and analytics: From big data to big impact. *MIS Quarterly*, 36(4):1165–1188, 2012.

[70] Shuqin Chen, Weiwei Yang, Hiroshi Yoshino, Mark D Levine, Katy Newhouse, and Adam Hinge. Definition of occupant behavior in residential buildings and its application to behavior analysis in case studies. *Energy and Buildings*, 104:1–13, 2015.

[71] X. W. Chen and X. Lin. Big data deep learning: Challenges and perspectives. *IEEE Access*, 2:514–525, 2014.

[72] Y. Chen, A. Ganapathi andR. Griffith, and R. Kat. The Case for Evaluating MapReduce Performance Using Workload Suites. In *2011 IEEE 19th Annual International Symposium on Modelling, Analysis, and Simulation of Computer and Telecommunication Systems*, pages 390–399. IEEE, 2011.

[73] Stefano Chessa, Michele Girolami, Luca Foschini, Raffaele Ianniello, Antonio Corradi, and Paolo Bellavista. Mobile crowd sensing management with the ParticipAct living lab. *Pervasive and Mobile Computing*, 38:200 – 214, 2017.

[74] Kyunghyun Cho, Bart van Merrienboer, Caglar Gulcehre, Dzmitry Bahdanau, Fethi Bougares, Holger Schwenk, and Yoshua Bengio. Learning phrase representations using RNN encoder–decoder for statistical machine translation. In *Conference on Empirical Methods in Natural Language Processing (EMNLP)*, pages 1724–1734. Association for Computational Linguistics, 2014.

[75] Chidchanok Choksuchat and Chantana Chantrapornchai. Benchmarking query complexity between rdb and owl. In *International Conference on Future Generation Information Technology*, pages 352–364. Springer, 2010.

[76] Peter Christoff. Renegotiating nature in the anthropocene: Australia's environment movement in a time of crisis. *Environmental Politics*, 25(6):1034–1057, 2016.

[77] T. Cioara, I. Anghel, I. Salomie, L. Barakat, S. Miles, D. Reidlinger, A. Taweel, C. Dobre, and F. Pop. Expert system for nutrition care process of older adults. *Future Generation Computer Systems*, 80:368–383, 2018.

[78] T. Cioara, I. Anghel, I. Salomie, and M. Dinsoreanu. A context - based semantically enhanced information retrieval model. In *IEEE 5th International Conference on Intelligent Computer Communication and Processing*, pages 245–250. IEEE, 2009.

[79] T. Cioara, I. Anghel, I. Salomie, and M. Dinsoreanu. A Policy-Based Context Aware Self-Management Model. In *11th International Symposium on Symbolic and Numeric Algorithms for Scientific Computing*, pages 333–340. IEEE, 2009.

[80] Smart city cluster Task 1. Website: `http://www.dolfin-fp7.eu/wp-content/uploads/2014/01/Task-1-List-of-DC-Energy-Related-Metrics-Final.pdf`.

[81] Dan Cole and Greg Siepen. Considerations for large-scale biodiversity reforestation plantings. part 8: Project monitoring and evaluation. *Australasian Plant Conservation: Journal of the Australian Network for Plant Conservation*, 25(1):17, 2016.

[82] Andrew Cook, Göksel Mısırlı, and Zhong Fan. Anomaly detection for iot time-series data: A survey. *IEEE Internet of Things Journal*, 2020.

[83] European Commission CORDIS. Caenti coordination action of the european network of territorial intelligence.

[84] E. Costanzo. *Wider use of EPBD databases- Enabling monitoring and policy making. EU CA EPBD factsheet*, 2019 (accessed February 2020).

[85] Ezilda Costanzo, David Weatherall, Andreas Androutsopoulos, and Virginia Gomez Onate. Can big data drive the market for residential energy efficiency. *ECEEE Summer Study Proc*, 1869, 2018.

[86] JP. Courtier. A strategy for operational implementation of 4d-var, using an incremental approach. *Q J R Meteorol Soc*, 120(519):1367–1387, 1994.

[87] Christophe Cruz. Semantic trajectory modeling for dynamic built environments. In *2017 IEEE International Conference on Data Science and Advanced Analytics (DSAA)*, pages 468–476. IEEE, 2017.

[88] D2RQ platform. Website: http://d2rq.org/.

[89] A. Daouadji, K.-K. Nguyen, M. Lemay, and M. Cheriet. Ontology-Based Resource Description and Discovery Framework for Low Carbon Grid Networks. In *2010 First IEEE International Conference on Smart Grid Communications*, pages 477–482. IEEE, 2010.

[90] Jason V. Davis, Brian Kulis, Prateek Jain, Suvrit Sra, and Inderjit S. Dhillon. Information-theoretic metric learning. In *International Conference on Machine Learning*, pages 209–216. ACM Press, 2007.

[91] M. Deakin. *Smart Cities: Governing, modelling and Analysing the transition*. Routledge, 8 2013.

[92] BIM Definition. Frequently asked questions about the national bim standard-united states-national bim standard-united states. *Nationalbimstandard. org. Archived from the original on*, 16, 2014.

[93] Presidenza del Consiglio dei Ministri. Dipartimento della protezione civile.

[94] R. Deng, R. Lu, C. Lai, T. H. Luan, and H. Liang. Optimal workload allocation in fog-cloud computing toward balanced delay and power consumption. *IEEE Internet of Things Journal*, 3(6):1171–1181, Dec 2016.

[95] Y. Deng, R. Sarkar, H. Ramasamy, R. Hosn, and R. Mahindru. An Ontology-Based Framework for Model-Driven Analysis of Situations in Data Centers. In *2013 IEEE International Conference on Services Computing*, pages 288–295. IEEE, 2013.

[96] Alex Densmore and Moin Hanif. *Modeling the condition of lithium ion batteries using the extreme learning machine*, pages 184–188. 2016.

[97] Andy Dent. *Getting started with LevelDB*. Packt Publishing Ltd, 2013.

[98] DIPENDE. Database integrato per la pianificazione energetica dei distretti edilizi, 2016 (accessed February 2020).

[99] P. Dumas, J.-P. Gardere, and Y. Bertacchini. Contribution of socio-technical systems theory concepts to a framework of territorial intelligence. In Blanca Miedes Ugarte Jean Jacques Girardot, editor, *Procs of International Conference of Territorial Intelligence.*, pages 92–105. Observatorio Local de Empleo, 2007.

[100] Ted Dunning and Ellen Friedman. *Time Series Databases: New Ways to Store and Access Data*. O'Reilly Media, Inc., 2014.

[101] L D'Amore, R Arcucci, L Marcellino, and A Murli. A parallel three-dimensional variational data assimilation scheme. In *AIP Conference Proceedings*, volume 1389, pages 1829–1831. AIP, 2011.

[102] Pebesma E. and Bivand R. Spatial data science.

[103] Greene K. Eagle N. *Reality Mining Using Big Data to Engineer a Better World*. MIT Press Ltd, 2014.

[104] U.K. EGHAM. Gartner says 5.8 billion enterprise and automotive iot endpoints will be in use in 2020. https://www.gartner.com/en/newsroom/press-releases/2019-08-29-gartner-says-5-8-billion-enterprise-and-automotive-io, 2019. (accessed January 16, 2020).

[105] A. Eldawy. SpatialHadoop: Towards flexible and scalable processing using MapReduce. 2014.

[106] A. Eldawy and M. F. Mokbel. Spatialhadoop: A mapreduce framework for spatial data. In *2015 IEEE 31st International Conference on Data Engineering, ICDE 2015*, Proceedings - International Conference on Data Engineering, pages 1352–1363. IEEE, apr 2015. 2015 31st IEEE International Conference on Data Engineering, ICDE 2015 ; Conference date: 13-04-2015 Through 17-04-2015.

[107] Kailas Elekar, MM Waghmare, and Amrit Priyadarshi. Use of rule base data mining algorithm for intrusion detection. In *2015 International Conference on Pervasive Computing (ICPC)*, pages 1–5. IEEE, 2015.

[108] Kailas Shivshankar Elekar. Combination of data mining techniques for intrusion detection system. In *2015 International Conference on Computer, Communication and Control (IC4)*, pages 1–5, 2015.

[109] Jeffrey L Elman. Finding structure in time. *Cognitive science*, 14(2):179–211, 1990.

[110] Jeffrey L. Elman. Finding structure in time. *Cognitive Science*, 14(2):179–211, mar 1990.

[111] embitel. How HARA Helps Functional Safety (ISO 26262) Consultants to Determine ASIL Values and Formulate Safety Goals. https://www.embitel.com/blog/embedded-blog/hara-by-iso-26262-standard-for-your-functional-safety-project, 2019. (accessed January 25, 2020).

[112] European Commission Environment. The seveso directive - technological disaster risk reduction.

[113] Lollini. R et al. *EXCEED Guidelines project findings and technical solutions. In exCEED - Knowledge sharing on buildings data*, 2019 (accessed February 2020).

[114] EU. Eu building stock observatory, 2016 (accessed February 2020).

[115] M. R. Evans, D. Oliver, X. Zhou, and S. Shekhar. *Spatial big data: Case studies on volume, velocity, and variety*. CRC Press, 2014.

[116] Franz Faerber, Alfons Kemper, Per Åke Larson, Justin Levandoski, Thomas Neumann, and Andrew Pavlo. Main memory database systems. *Foundations and Trends® in Databases*, 8(1-2):1–130, 2017.

[117] Zhun Fan, Youxiang Zuo, Fang Li, and Shuangxi Wang. Design of a diamond adsorption detection system based on machine learning techniques. In *2016 12th World Congress on Intelligent Control and Automation (WCICA)*, pages 3124–3128, 2016.

[118] Bronwyn A. Fancourt, Brooke L. Bateman, Jeremy VanDerWal, Stewart C. Nicol, Clare E. Hawkins, Menna E. Jones, and Christopher N. Johnson. Testing the role of climate change in species decline: Is the eastern quoll a victim of a change in the weather? *PLOS ONE*, 10(6):e0129420, 2015.

[119] C. Fiandrino, N. Allio, D. Kliazovich, P. Giaccone, and P. Bouvry. Profiling performance of application partitioning for wearable devices in mobile cloud and fog computing. *IEEE Access*, 7:12156–12166, Jan 2019.

[120] C. Fiandrino, A. Capponi, G. Cacciatore, D. Kliazovich, U. Sorger, P. Bouvry, B. Kantarci, F. Granelli, and S. Giordano. CrowdSenSim: a simulation platform for mobile crowdsensing in realistic urban environments. *IEEE Access*, 5:3490–3503, Feb 2017.

[121] C. Fiandrino, D. Kliazovich, P. Bouvry, and A. Y. Zomaya. Performance and energy efficiency metrics for communication systems of cloud computing data centers. *IEEE Transactions on Cloud Computing*, 5(4):738–750, Oct 2017.

[122] C. Fiandrino, A. Blanco Pizarro, P. Jiménez Mateo, C. Andrés Ramiro, N. Ludant, and J. Widmer. openLEON: An end-to-end emulation platform from the edge data center to the mobile user. *Computer Communications*, 148:17 – 26, 2019.

[123] Miltiades C Filippou, Dario Sabella, and Vincenzo Riccobene. Flexible MEC service consumption through edge host zoning in 5G networks. *CoRR*, abs/1903.01794, 2019.

[124] M. M. Fischer. From conventional to knowledge-based geographic information systems. *Computers, Environment and Urban Systems*, 18(4):233–242, 7 1994. Special Issue: 16th Urban Data Management Symposium.

[125] D. Fitzner, M. Sester, U. Haberlandt, and E. Rabiei. Rainfall estimation with a geosensor network of cars – theoretical considerations and first results. *Photogrammetrie - Fernerkundung - Geoinformation*, 2013:93–103, 05 2013.

[126] R. Ford, C. C. Pain, A. J. H. Goddard, C. R. E. De Oliveira, and A. P. Umpleby. A non-hydrostatic finite-element model for three-dimensional stratified oceanic flows. part I: Model formulation. *Monthly Weather Review*, 132:2816–2831, 2004.

[127] Piotr Fryzlewicz. Wild binary segmentation for multiple change-point detection. *The Annals of Statistics*, 42(6):2243–2281, 2014.

[128] A. Furno, M. Fiore, and R. Stanica. Joint spatial and temporal classification of mobile traffic demands. In *Proc. of IEEE INFOCOM*, pages 1–9, May 2017.

[129] Angelo Furno, Marco Fiore, Razvan Stanica, Cezary Ziemlicki, and Zbigniew Smoreda. A tale of ten cities: Characterizing signatures of mobile traffic in urban areas. *IEEE Transactions on Mobile Computing*, 16(10):2682–2696, 2017.

[130] M. Gahegan. Fourth paradigm giscience? prospects for automated discovery and explanation from data. *International Journal of Geographical Information Science*, 34(1):1–21, 9 2019.

[131] M. Garschagen. *Risky change? Vulnerability and adaptation between climate change and transformation dynamics in Can Tho City, Vietnam.* Franz Steiner Verlag.

[132] Julien Gedeon, Jeff Krisztinkovics, Christian Meurisch, Michael Stein, Lin Wang, and Max Mühlhäuser. A multi-cloudlet infrastructure for future smart cities: An empirical study. In *Proc. ACM EdgeSys*, pages 19–24, Jun 2018.

[133] R. Gerundo, I. Fasolino, and M. Grimaldi. *ISUT Model. A Composite Index to Measure the Sustainability of the Urban Transformation*, page 117–130. Springer International Publishing, 2016.

[134] R. Gibbons. *A primer in Game Theory.* Prentice Hall Inc, USA, 1992.

[135] A. Ginige, L. Paolino, M. Romano, M. Sebillo, G. Tortora, and G. Vitiello. Information sharing among disaster responders - an interactive spreadsheet-based collaboration approach. *Computer Supported Cooperative Work (CSCW)*, 23(4-6):547–583, jul 2014.

[136] J. J. Girardot and E. Brunau. Inteligencia territorial e innovación para el desarrollo socio-ecológica transición. ecological and social innovation. In Blanca Miedes Ugarte Jean Jacques Girardot, editor, *Procs of International Conference of Territorial Intelligence.*, pages 6–6. INTI-International Network of Territorial Intelligence.

[137] Fabio Giust, Gianluca Verin, Kiril Antevski, Joey Chou, Yonggang Fang, Walter Featherstone, and et al. MEC deployments in 4G and evolution towards 5G, Feb 2018. ETSI White Paper.

[138] Marta C Gonzalez, Cesar A Hidalgo, and Albert-Laszlo Barabasi. Understanding individual human mobility patterns. *Nature*, 453(7196):779, 2008.

[139] Ian Goodfellow, Yoshua Bengio, and Aaron Courville. *Deep learning.* MIT press, 2016.

[140] J. P. Gouveia, G. Giannakidis, and J. Seixas. *Smart City Energy Planning: Integrating Data and Tools*, 2018 (accessed February 2020).

[141] F. Granelli, R. Bassoli, and M. Di Renzo. Energy-efficiency analysis of cloud radio access network in heterogeneous 5G networks. In *Proc. of European Wireless Conference*, pages 1–6, May 2018.

[142] M. Grimaldi, M. Sebillo, G. Vitiello, and V. Pellecchia. *An Ontology Based Approach for Data Model Construction Supporting the Management and Planning of the Integrated Water Service*, page 243–252. Springer International Publishing, 2019.

[143] Katarina Grolinger, Wilson A Higashino, Abhinav Tiwari, and Miriam AM Capretz. Data management in cloud environments: Nosql and newsql data stores. *Journal of Cloud Computing: advances, systems and applications*, 2(1):22, 2013.

[144] Amit Gupta, Azeem Mohammad, Ali Syed, and Malka N. A comparative study of classification algorithms using data mining: Crime and accidents in denver city the usa. *International Journal of Advanced Computer Science and Applications*, 7(7), 2016.

[145] Harshit Gupta, Amir Vahid Dastjerdi, Soumya K Ghosh, and Rajkumar Buyya. iFogSim: A toolkit for modeling and simulation of resource management techniques in the internet of things, edge and fog computing environments. *Software: Practice and Experience*, 47(9):1275–1296, 2017.

[146] Manish Gupta, Jing Gao, Charu C. Aggarwal, and Jiawei Han. Outlier detection for temporal data: A survey. *IEEE Transactions on Knowledge and Data Engineering*, 26(9):2250–2267, sep 2014.

[147] Tomasz Górecki, Lajos Horváth, and Piotr Kokoszka. Change point detection in heteroscedastic time series. *Econometrics and Statistics*, 7:63–88, 2018.

[148] P. Haase, T. Mathaß, M. Schmidt andA. Eberhart, and U. Walther. Semantic Technologies for Enterprise Cloud Management. In *International Semantic Web Conference ISWC 2010*, volume 6497, pages 98–113. Springer, 2010.

[149] Malka N Halgamuge, Siddeswara M Guru, and Andrew Jennings. *Centralised strategies for cluster formation in sensor networks.* Springer, 2005.

[150] Malka N Halgamuge, Siddeswara Mayura Guru, and Andrew Jennings. Energy efficient cluster formation in wireless sensor networks. In *10th International Conference on Telecommunications, 2003. ICT 2003*, pages 1571–1576, 2003.

[151] Chulwoo Han and Abderrahim Taamouti. Partial structural break identification. *Oxford Bulletin of Economics and Statistics*, 79(2):145–164, 2017.

[152] A. Hannachi. A primer for eof analysis of climate data. *Department of Meteorology, University of Reading, UK*, 2004.

[153] A. Hannachi, I.T. Jolliffe, and D.B. Stephenson. Empirical orthogonal functions and related techniques in atmospheric science: A review. *International Journal of Climatology: A Journal of the Royal Meteorological Society*, 27:1119–1152, 2007.

[154] P. C. Hansen, J. G. Nagy, and D. P. O'Leary. Deblurring images: Matrices, spectra, and filtering. *SIAM*, 2006.

[155] Zaid Harchaoui and Olivier Cappe. Retrospective mutiple change-point estimation with kernels. In *IEEE Workshop on Statistical Signal Processing*, pages 768–772. IEEE, 2007.

[156] IPDJ Harris, Philip D Jones, Timothy J Osborn, and David H Lister. Updated high-resolution grids of monthly climatic observations–the cru ts3. 10 dataset. *International journal of climatology*, 34(3):623–642, 2014.

[157] J. Hasenburg, M. Grambow, E. Grünewald, S.Huk, and D. Bermbach. MockFog: Emulating fog computing infrastructure in the cloud. In *Accepted in Proc. IEEE International Conference on Fog Computing*, pages 1–6, June 2019.

[158] Ying He, F. Richard Yu, Nan Zhao, Victor C. M. Leung, and Hongxi Yin. Software-defined networks with mobile edge computing and caching for smart cities: A big data deep reinforcement learning approach. *IEEE Communications Magazine*, 55(12):31–37, dec 2017.

[159] John Hebeler, Matthew Fisher, Ryan Blace, and Andrew Perez-Lopez. *Semantic Web Programming.* John Wiley & Sons, 2009.

[160] Hansika Hewamalage, Christoph Bergmeir, and Kasun Bandara. Recurrent neural networks for time series forecasting: Current status and future directions, 2019.

[161] Hitachi. Geomondrian project.

[162] P. Hitzler, K. Markus, and R. Sebastian. *Foundations of Semantic Web Technologies.* Chapman and Hall/CRC Press, 2010.

[163] Sepp Hochreiter and Jürgen Schmidhuber. Long short-term memory. *Neural Computation*, 9(8):1735–1780, nov 1997.

[164] Tianzhen Hong, Simona D'Oca, William JN Turner, and Sarah C Taylor-Lange. An ontology to represent energy-related occupant behavior in buildings. part i: Introduction to the dnas framework. *Building and Environment*, 92:764–777, 2015.

[165] Mohammad Shahadat Hossain, Faisal Ahmed, Karl Andersson, et al. A belief rule based expert system to assess tuberculosis under uncertainty. *Journal of medical systems*, 41(3):43, 2017.

[166] Mohammad Shahadat Hossain, Israt Binteh Habib, and Karl Andersson. A belief rule based expert system to diagnose dengue fever under uncertainty. In *2017 Computing Conference*, pages 179–186. IEEE, 2017.

[167] Mohammad Shahadat Hossain, Md Saifuddin Khalid, Shamima Akter, and Shati Dey. A belief rule-based expert system to diagnose influenza. In *2014 9Th international forum on strategic technology (IFOST)*, pages 113–116. IEEE, 2014.

[168] Mohammad Shahadat Hossain, Saifur Rahaman, Ah-Lian Kor, Karl Andersson, and Colin Pattinson. A belief rule based expert system for datacenter pue prediction under uncertainty. *IEEE Transactions on Sustainable Computing*, 2(2):140–153, 2017.

[169] Mohammad Shahadat Hossain, Pär-Ola Zander, Md Sarwar Kamal, and Linkon Chowdhury. Belief-rule-based expert systems for evaluation of e-government: a case study. *Expert Systems*, 32(5):563–577, 2015.

[170] Rob J. Hyndman and George Athanasopoulos. *Forecasting: principles and practice.* OTexts, 2018.

[171] IEC 62443-3-3: Industrial communication networks – network and system security – part 3-3: System security requirements and security levels, 2013.

[172] InfluxData. Influxdb documentation v1.8 https://docs.influxdata.com/influxdb/.

[173] Territorial Intelligence. European network of territorial intelligence (enti), 2018.

[174] ISO 15408, information technology - security techniques - evaluation criteria for IT security (Common Criteria), 2009.

[175] Road vehicles-functional safety-part 2: Management of functional safety, ISO 26262, 2018.

[176] ISO/SAE DIS 21434. Road vehicles – cybersecurity engineering. https://www.iso.org/standard/70918.html, 2020. (accessed February 16, 2020).

[177] R. Buyya JA. Khosravi. Energy and Carbon Footprint-Aware Management of Geo-Distributed Cloud Data Centers: A Taxonomy. *State of the Art, and Future Directions*, 2018.

[178] Hyeon Ae Jang, SH Hong, and MK Lee. A study on situation analysis for asil determination. *Journal of Industrial and Intelligent Information*, 3(2), 2015.

[179] Kasthuri Jayarajah, Andrew Tan, and Archan Misra. Understanding the interdependency of land use and mobility for urban planning. In *Proc. ACM UbiComp*, pages 1079–1087, 2018.

[180] C. S. Jayasekara, M. N Halgamuge, A Noor, and A Saeed. *Analysis of traffic offenses in transportation: Application of Big Data Analysis*. CRC Press, 2018.

[181] Jena tool. Website: `https://jena.apache.org/`.

[182] M. Jia, J. Cao, and W. Liang. Optimal cloudlet placement and user to cloudlet allocation in wireless metropolitan area networks. *IEEE Transactions on Cloud Computing*, 5(4):725–737, Oct 2017.

[183] Mengda Jia, Ali Komeily, Yueren Wang, and Ravi S Srinivasan. Adopting internet of things for the development of smart buildings: A review of enabling technologies and applications. *Automation in Construction*, 101:111–126, 2019.

[184] Z. Jiang and S. Shekhar. *Spatial Big Data Science - Classification Techniques for Earth Observation Imagery*. Springer, 2017.

[185] John W Jones. Efficient wetland surface water detection and monitoring via landsat: Comparison with in situ data from the everglades depth estimation network. *Remote Sensing*, 7(9):12503–12538, 2015.

[186] John W Jones. Improved automated detection of subpixel-scale inundation—revised dynamic surface water extent (dswe) partial surface water tests. *Remote Sensing*, 11(4):374, 2019.

[187] Michael I Jordan. Serial order: A parallel distributed processing approach. In *Advances in psychology*, volume 121, pages 471–495. Elsevier, 1997.

[188] Josverwoerd. Digesting big data. https://blog.bigml.com/2012/11/12/digesting-big-data/, 2012. Accessed on: 23-01-2020.

[189] N. Jukic, B. Jukic, and M. Malliaris. *Online Analytical Processing (OLAP) for Decision Support*, page 259–276. Springer Berlin Heidelberg, 2008.

[190] Mahmood Hussain Kadhem and Ahmed M Zeki. Prediction of urinary system disease diagnosis: A comparative study of three decision tree algorithms. In *2014 International Conference on Computer Assisted System in Health*, pages 58–61. IEEE, 2014.

[191] Henning Kagermann, Johannes Helbig, Ariane Hellinger, and Wolfgang Wahlster. *Recommendations for implementing the strategic initiative INDUSTRIE 4.0: Securing the future of German manufacturing industry; final report of the Industrie 4.0 Working Group*. Forschungsunion, 2013.

[192] R.E. Kalman. A new approach to linear filtering and prediction problems. *Trans. ASME J. Basic Eng.*, 82(Series D):35–45, 1960.

[193] E. Kalnay. *Atmospheric modeling, data assimilation and predictability*. Cambridge, 2003.

[194] Ali Asgher Kapadiya. Visualisation of multi-service system net-work with d3.js & kdb+/q using websocket. *Global Journal of Computer Science and Technology*, 2018.

[195] Razuan Karim, Karl Andersson, Mohammad Shahadat Hossain, Md Jasim Uddin, and Md Perveg Meah. A belief rule based expert system to assess clinical bronchopneumonia suspicion. In *2016 Future Technologies Conference (FTC)*, pages 655–660. IEEE, 2016.

[196] H. A. Karimi and B. Karimi. *Geospatial Data Science Techniques and Applications.* CRC Press, 2017.

[197] MARTIN KASTEBO and VICTOR NORDH. Model-based security testing in automotive industry. Master's thesis, Department of Computer Science and Engineering - UNIVERSITY OF GOTHENBURG, Gothenburg, Sweden, 2017.

[198] D. Keim, G. Andrienko, J.-D. Fekete, C. Görg, J. Kohlhammer, and G. Melançon. *Visual Analytics: Definition, Process, and Challenges*, page 154–175. Springer Berlin Heidelberg.

[199] E. Keogh, S. Chu, D. Hart, and M. Pazzani. An online algorithm for segmenting time series. In *IEEE International Conference on Data Mining*, pages 289–296. IEEE Comput. Soc, 2001.

[200] R. Killick, P. Fearnhead, and I. A. Eckley. Optimal detection of changepoints with a linear computational cost. *Journal of the American Statistical Association*, 107(500):1590–1598, 2012.

[201] Hyeon Gyu Kim, Yoo Hyun Park, Yang Hyun Cho, and Myoung Ho Kim. Time-slide window join over data streams. *Journal of Intelligent Information Systems*, 43(2):323–347, 2014.

[202] Maral Kolahkaj and Madjid Khalilian. A recommender system by using classification based on frequent pattern mining and j48 algorithm. In *2015 2nd International Conference on Knowledge-Based Engineering and Innovation (KBEI)*, pages 780–786, 2015.

[203] Philip Koopman. Practical experience report: Automotive safety practices vs. accepted principles. In *International Conference on Computer Safety, Reliability, and Security*, pages 3–11. Springer, 2018.

[204] Karl Koscher, Alexei Czeskis, Franziska Roesner, Shwetak Patel, Tadayoshi Kohno, Stephen Checkoway, Damon McCoy, Brian Kantor, Danny Anderson, Hovav Shacham, et al. Experimental security analysis of a modern automobile. In *2010 IEEE Symposium on Security and Privacy*, pages 447–462. IEEE, 2010.

[205] Limin Kou, Xiangyang Li, Jianyi Lin, and Jiefeng Kang. Simulation of urban water resources in xiamen based on a weap model. *Water*, 10(6):732, 2018.

[206] Mamoru Kubo, Shu Nishikawa, Eiji Yamamoto, and Ken-ichiro Muramoto. Identification of individual tree crowns from satellite image and image-to-map rectification. In *2007 IEEE International Geoscience and Remote Sensing Symposium*, pages 1905–1908, 2007.

[207] Kx Labs. Kdb+ documentation v3.6 https://kx.com/kdb-documentation/.

[208] D. Laney. Deja vvvu: Others claiming gartner's construct for big data.

[209] Bob Lantz, Brandon Heller, and Nick McKeown. A network in a laptop: Rapid prototyping for software-defined networks. In *Proc. of ACM Hotnets-IX*, pages 1–6, 2010.

[210] R. Laurini. *Geographic Knowledge Infrastructure. Applications to Territorial Intelligence and Smart Cities.*

[211] R. Laurini. *Nature of Geographic Knowledge Bases*, pages 29–60. IGI Global, 2017.

[212] Sanja Lazarova-Molnar and Nader Mohamed. Collaborative data analytics for smart buildings: opportunities and models. *Cluster Computing*, 22(1):1065–1077, 2019.

[213] Data Learning. https://www.imperial.ac.uk/data-science/research/research-themes/datalearning/.

[214] Y. LeCun, Y. Bengio, and G. Hinton. Deep learning. *Nature*, 521(7553):436–444, 5 2015.

[215] J.-G. Lee, J. Han, X. Li, and H. Gonzalez. TraClass: trajectory classification using hierarchical region-based and trajectory-based clustering. *Proc. VLDB Endow.*, 1(1):1081–1094, 2008.

[216] L. Li, W. Zheng, X. Wang, and X. Wang. Coordinating liquid and free air cooling with workload allocation for data center power minimization. In *11th International Conference on Autonomic Computing (ICAC)*, pages 249–259. USENIX Association, 2014.

[217] S. Li, S. Dragicevic, F. A. Castro, M. Sester, S. Winter, A. Coltekin, C. Pettit, B. Jiang, J. Haworth, A. Stein, and et al. Geospatial big data handling theory and methods: A review and research challenges. *ISPRS Journal of Photogrammetry and Remote Sensing*, 115:119–133, May 2016.

[218] J. Liang, Z. Chen, C. Li, and B. Xia. Delay outage probability of multi-relay selection for mobile relay edge computing system. pages 898–902, Aug 2019.

[219] Jing J Liang, A Kai Qin, Ponnuthurai N Suganthan, and S Baskar. Comprehensive learning particle swarm optimizer for global optimization of multimodal functions. *IEEE transactions on evolutionary computation*, 10(3):281–295, 2006.

[220] JJ Liang, BY Qu, PN Suganthan, and Alfredo G Hernández-Díaz. Problem definitions and evaluation criteria for the cec 2013 special session on real-parameter optimization. *Computational Intelligence Laboratory, Zhengzhou University, Zhengzhou, China and Nanyang Technological University, Singapore, Technical Report*, 201212(34):281–295, 2013.

[221] Edward M Lim, Miguel Molina Solana, Christopher Pain, Yi-Ke Guo, and Rossella Arcucci. Hybrid data assimilation: An ensemble-variational approach. In *2019 15th International Conference on Signal-Image Technology & Internet-Based Systems (SITIS)*, pages 633–640. IEEE, 2019.

[222] Isis Didier Lins, Moacyr Araujo, Márcio das Chagas Moura, Marcus André Silva, and Enrique López Droguett. Prediction of sea surface temperature in the tropical atlantic by support vector machines. *Computational Statistics & Data Analysis*, 61:187–198, 2013.

[223] Jianxiao Liu, Zonglin Tian, Panbiao Liu, Jiawei Jiang, and Zhao Li. An approach of semantic web service classification based on naive bayes. In *2016 IEEE International Conference on Services Computing (SCC)*, pages 356–362, 2016.

[224] P. Liu, D. Willis, and S. Banerjee. ParaDrop: Enabling lightweight multi-tenancy at the network's extreme edge. In *Proc. IEEE/ACM SEC*, pages 1–13, Oct 2016.

[225] Siqi Liu, Adam Wright, and Milos Hauskrecht. Change-point detection method for clinical decision support system rule monitoring. *Artificial Intelligence in Medicine*, 91:49–56, 2018.

[226] Xiaochen Liu, Shupeng Sun, Xin Li, Haifeng Qian, and Pingqiang Zhou. Machine learning for noise sensor placement and full-chip voltage emergency detection. *IEEE Transactions on Computer-Aided Design of Integrated Circuits and Systems*, pages 1–1, 2016.

[227] Z. Liu, I. Liu, S. Low, and A. Wierman. Pricing Data Center Demand Response. In *SIG-METRICS '14: The 2014 ACM International Conference on Measurement and Modeling of Computer Systems*, page 111–123. ACM, 2014.

[228] A.C. Lorenc. Development of an operational variational assimilation scheme. *Journal of the Meteorological Society of Japan*, 75:339–346, 1997.

[229] E.N. Lorenz. Empirical orthogonal functions and statistical weather prediction., 1956.

[230] Josip Lorincz, Tonko Garma, and Goran Petrovic. Measurements and modelling of base station power consumption under real traffic loads. In *Sensors*, 2012.

[231] P. Louridas and C. Ebert. Machine learning. *IEEE Software*, 33(5):110–115, 2016.

[232] Lucid and Urjanet. *eBook / Practical Guide to Transforming Energy Data into Better Buildings.US RICS, BSO 2nd Stakeholder Workshop*,, 2019 (accessed February 2020).

[233] Alexandre Lung-Yut-Fong, Céline Lévy-Leduc, and Olivier Cappé. Homogeneity and change-point detection tests for multivariate data using rank statistics. *Journal de la Société Française de Statistique*, 156(4):133–162, 2015.

[234] Guofeng Ma and Zhijiang Wu. Bim-based building fire emergency management: Combining building users' behavior decisions. *Automation in Construction*, 109:102975, 2020.

[235] Zhendong Ma, Aleksandar Hudic, Abdelkader Shaaban, and Sandor Plosz. Security viewpoint in a reference architecture model for cyber-physical production systems. In *2017 IEEE European Symposium on Security and Privacy Workshops (EuroS&PW)*, pages 153–159. IEEE, 2017.

[236] A. M. MacEachren. *Leveraging Big (Geo) Data with (Geo) Visual Analytics: Place as the Next Frontier*, pages 139–155. Springer Singapore, 2017.

[237] Georg Macher, Eric Armengaud, Eugen Brenner, and Christian Kreiner. Threat and risk assessment methodologies in the automotive domain. *Procedia computer science*, 83:1288–1294, 2016.

[238] A. A Raneesha Madushanki, Malka N Halgamuge, W. A. H. S Wirasagoda, and Ali Syed. Adoption of the internet of things (iot) in agriculture and smart farming towards urban greening: A review. *International Journal of Advanced Computer Science and Applications*, 10(4), 2019.

[239] Hardik H Maheta and Vipul K Dabhi. *Classification of imbalanced data sets using multi objective genetic programming*, pages 1–6. 2015.

[240] Tanjim Mahmud and Mohammad Shahadat Hossain. An evidential reasoning-based decision support system to support house hunting. *International Journal of Computer Applications*, 57(21):51–58, 2012.

[241] Tanjim Mahmud, Kazi Namirur Rahman, and Mohammad Shahadat Hossain. Evaluation of job offers using the evidential reasoning approach. *Global Journal of Computer Science and Technology*, 2013.

[242] U. Mandal, M. F. Habib, S. Zhang, B. Mukherjee, and M. Tornatore. Greening the Cloud Using Renewable-Energy-Aware Service Migration. *IEEE Network*, 27(6):36–43, 2013.

[243] J. Manyika, M. Chui, B. Brown, J. Bughin, R. Dobbs, C. Roxburgh, and A. H. Byers. Big data: the next frontier for innovation, competition and productivity.

[244] Andrew R. Marshall, Philip J. Platts, Roy E. Gereau, William Kindeketa, Simon Kang'ethe, and Rob Marchant. The genus acacia (fabaceae) in east africa: distribution, diversity and the protected area network. *Plant Ecology and Evolution*, 145(3):289–301, 2012.

[245] R. Mayer, L. Graser, H. Gupta, E. Saurez, and U. Ramachandran. EmuFog: Extensible and scalable emulation of large-scale fog computing infrastructures. In *Proc. IEEE Fog World Congress*, pages 1–6, Oct 2017.

[246] McAfee. Automotive security best practices. Technical report, McAfee, June 2016. Recommendations for security and privacy in the era of the next-generation car.

[247] Katrina A McDonnell and Neil J Holbrook. A poisson regression model of tropical cyclogenesis for the australian–southwest pacific ocean region. *Weather and Forecasting*, 19(2):440–455, 2004.

[248] R. McFarlane. *A survey of Exploration Strategies in Reinforcement Learning*. McGill University, Scool of Computer Science, 1998.

[249] A. Meisner, G. B. De Deyn, W. de Boer, and W. H. van der Putten. Soil biotic legacy effects of extreme weather events influence plant invasiveness. *Proceedings of the National Academy of Sciences*, 110(24):9835–9838, 2013.

[250] Mesfin M Mekonnen and Arjen Y Hoekstra. Four billion people facing severe water scarcity. *Science advances*, 2(2):e1500323, 2016.

[251] Ryan Melfi, Ben Rosenblum, Bruce Nordman, and Ken Christensen. Measuring building occupancy using existing network infrastructure. In *2011 International Green Computing Conference and Workshops*, pages 1–8. IEEE, 2011.

[252] Efren Mezura-Montes, Margarita Reyes-Sierra, and Carlos A Coello Coello. Multi-objective optimization using differential evolution: a survey of the state-of-the-art. In *Advances in differential evolution*, pages 173–196. Springer, 2008.

[253] B. Miedes Ugarte. Territorial intelligence and the three components of territorial governance. In Blanca Miedes Ugarte Jean Jacques Girardot, editor, *Conference of Territorial Intelligence.*, page 10. Observatorio Local de Empleo.

[254] Charlie Miller and Chris Valasek. Remote exploitation of an unaltered passenger vehicle. *Black Hat USA*, 2015:91, 2015.

[255] Ramakanta Mohanty and K Jhansi Rani. Application of computational intelligence to predict churn and non-churn of customers in indian telecommunication. In *2015 International Conference on Computational Intelligence and Communication Networks (CICN)*, pages 598–603, 2015.

[256] Dorin Moldovan, Marcel Antal, Dan Valea, Claudia Pop, Tudor Cioara, Ionut Anghel, and Ioan Salomie. Tools for mapping ontologies to relational databases: A comparative evaluation. In *2015 IEEE International Conference on Intelligent Computer Communication and Processing (ICCP)*, pages 77–83. IEEE, 2015.

[257] A. Monti, D. Pesch, K. Ellis, and P. Mancarella. *Energy Positive Neighborhoods and Smart Energy Districts: Methods, Tools, and Experiences from the Field*. Academic Press, 2017.

[258] Federico Montori, Emanuele Cortesi, Luca Bedogni, Andrea Capponi, Claudio Fiandrino, and Luciano Bononi. CrowdSenSim 2.0: A stateful simulation platform for mobile crowdsensing in smart cities. In *Proc. of ACM MSWIM*, page 289–296, New York, NY, USA, 2019.

[259] R. Morabito, V. Cozzolino, A. Y. Ding, N. Beijar, and J. Ott. Consolidate IoT edge computing with lightweight virtualization. *IEEE Network*, 32(1):102–111, Jan 2018.

[260] Jacob Morgan. A simple explanation of 'the internet of things'. https://www.forbes.com/sites/jacobmorgan/2014/05/13/simple-explanation-internet-things-that-anyone-can-understand/, may 2014. (accessed January 17, 2020).

[261] C. Mouradian, D. Naboulsi, S. Yangui, R. H. Glitho, M. J. Morrow, and P. A. Polakos. A comprehensive survey on fog computing: State-of-the-art and research challenges. *IEEE Communications Surveys Tutorials*, 20(1):416–464, First Quarter 2018.

[262] M. A. Musen. The protégé project. *AI Matters*, 1(4):4–12, Jun 2015.

[263] Ashish Nargundkar and YS Rao. Influencerank: A machine learning approach to measure influence of twitter users. In *2016 International Conference on Recent Trends in Information Technology (ICRTIT)*, pages 1–6, 2016.

[264] T.H. Naylor, T.H. Naylor, J.L. Balintfy, D.S. Burdick, and K. Chu. *Computer Simulation Techniques*. Wiley, 1966.

[265] Netscribes. The role of IoT in the future of the automotive industry. https://www.netscribes.com/the-present-and-future-role-of-automotive-iot/, 2018. (accessed January 25, 2020).

[266] NHTSA. Automated vehicles for safety. https://www.nhtsa.gov/technology-innovation/automated-vehicles-safety, 2017. (accessed February 10, 2020).

[267] N. Nichols. *Data Assimilation - Chapter Mathematical concepts in data assimilation*. Springer, 2010.

[268] Taher Niknam. A new fuzzy adaptive hybrid particle swarm optimization algorithm for non-linear, non-smooth and non-convex economic dispatch problem. *Applied Energy*, 87(1):327–339, 2010.

[269] Taher Niknam, Hassan Doagou Mojarrad, and Majid Nayeripour. A new fuzzy adaptive particle swarm optimization for non-smooth economic dispatch. *Energy*, 35(4):1764–1778, 2010.

[270] Nord Pool Spot Market Data. Website: `http://www.nordpoolspot.com/Market-data1/`.

[271] Alan T Norman. *Hacking: Computer Hacking Beginners Guide how to Hack Wireless Network, Basic Security and Penetration Testing, Kali Linux, Your First Hack*. CreateSpace Independent Publishing Platform, 2016.

[272] Juthamas Nuansanong, Supaporn Kiattisin, and Adisorn Leelasantitham. Diagnosis and interpretation of dental x-ray in case of deciduous tooth extraction decision in children using active contour model and j48 tree. In *2014 International Electrical Engineering Congress (iEECON)*, pages 1–4, 2014.

[273] Martin J O'Connor and Amar K Das. Sqwrl: A query language for owl. In *OWLED*, volume 529, 2009.

[274] OGC. The ogc's emerging role in geospatial business intelligence – geobi.

[275] Peter R Oke, Gary B Brassington, David A Griffin, and Andreas Schiller. The bluelink ocean data assimilation system (bodas). *Ocean Modelling*, 21(1-2):46–70, 2008.

[276] M Oliveira, A Grillo, and M Tabarelli. *Forest edge in the Brazilian Atlantic forest: drastic changes in tree species assemblages*. 2004.

[277] Ontotext. What are ontologies? https://ontotext.com/knowledgehub/fundamentals/what-are-ontologies/, 2018. (accessed January 26, 2020).

[278] OWL specification. Website: `http://www.w3.org/TR/owl2-overview/`.

[279] Rob O'Hara, S Green, and T McCarthy. The agricultural impact of the 2015–2016 floods in ireland as mapped through sentinel 1 satellite imagery. *Irish Journal of Agricultural and Food Research*, 58(1):44–65, 2019.

[280] D. J. P. Rashidi. *Handbook on Securing Cyber-Physical Critical Infrastructure*. 2012.

[281] A. Papoulis. *Probability, Random Variables, and Stochastic Processes*. McGraw-Hill International Editions, 1991.

[282] Tuomas Pelkonen, Scott Franklin, Justin Teller, Paul Cavallaro, Qi Huang, Justin Meza, and Kaushik Veeraraghavan. Gorilla: A fast, scalable, in-memory time series database. *Procedings of the VLDB Endowment*, 8(12):1816–1827, 2015.

[283] Pellet. Website: `http://clarkparsia.com/pellet/`.

[284] R. J. Pooley and P. Stevens. *Using UML: Software Engineering with Objects and Components*.

[285] Prometheus. Prometheus documentation v2.18.1 https://prometheus.io/docs/.

[286] Protege tool. Website: `http://protege.stanford.edu/`.

[287] Eetu Puttonen, Anttoni Jaakkola, Paula Litkey, and Juha Hyyppä. Tree classification with fused mobile laser scanning and hyperspectral data. *Sensors*, 11(5):5158–5182, 2011.

[288] T. Qayyum, A. W. Malik, M. A. Khan Khattak, O. Khalid, and S. U. Khan. Fognet-sim++: A toolkit for modeling and simulation of distributed fog environment. *IEEE Access*, 6:63570–63583, 2018.

[289] Zhongjun Qu and Pierre Perron. Estimating and testing structural changes in multivariate regressions. *Econometrica*, 75(2):459–502, 2007.

[290] César Quilodrán Casas. *Fast ocean data assimilation and forecasting using a neural-network reduced-space regional ocean model of the north Brazil current*. PhD thesis, Imperial College London, 2018.

[291] Rohini Rajpal, Sanmeet Kaur, and Ramandeep Kaur. Improving detection rate using misuse detection and machine learning. In *2016 SAI Computing Conference (SAI)*, pages 1131–1135, 2016.

[292] RAPID7. Vulnerabilities, exploits, and threats - defining three key terms in cybersecurity. https://www.rapid7.com/fundamentals/vulnerabilities-exploits-threats/, 2019. (accessed January 29, 2020).

[293] Asanga Ratnaweera, Saman K Halgamuge, and Harry C Watson. Self-organizing hierarchical particle swarm optimizer with time-varying acceleration coefficients. *IEEE Transactions on evolutionary computation*, 8(3):240–255, 2004.

[294] J Refonaa, M Lakshmi, and V Vivek. Analysis and prediction of natural disaster using spatial data mining technique. In *2015 International Conference on Circuits, Power and Computing Technologies [ICCPCT-2015]*, pages 1–6, 2015.

[295] Reticular Systems. Inc. Artificial Intelligence for Communications Resource Management.

[296] Guillem Rigaill. A pruned dynamic programming algorithm to recover the best segmentations with 1 to k_max change-points. *Journal de la Société Française de Statistique*, 156(4):180–205, 2015.

[297] A Robinson. Geovisual analytics. *The Geographic Information Science & Technology Body of Knowledge (3rd Quarter 2017 Edition), edited by John P. Wilson. UCGIS. https://doi. org/10.22224/gistbok/2017.3*, 6, 2017.

[298] V. P. Roijers D.M. A Survey of Multi-Objective Sequential Decision-Making. *Journal of Artificial Intelligence Research*, pages 67–113, 2013.

[299] Alessandra Tassab Roya Ayazia, Ilaria d'Auriaa and Julien Turpin. *The Ever Growing use of Copernicus across Europe's Regions: a selection of 99 user stories*, 2018 (accessed May 3, 2020).

[300] L. Y. Rusin, E. V. Lyubetskaya, K. Y. Gorbunov, and V. A. Lyubetsky. Reconciliation of gene and species trees. *BioMed Research International*, 2014:1–22, 2014.

[301] P D Rymer, M Sandiford, S A Harris, M R Billingham, and D H Boshier. Remnant pachira quinata pasture trees have greater opportunities to self and suffer reduced reproductive success due to inbreeding depression. *Heredity*, 115(2):115–124, 2013.

[302] Y. X. S. Yi. Cs2p: Improving video bitrate selection and adaptation with data-driven throughput prediction. *SIGCOMM'16. New York, USA: ACM*, 2016.

[303] Rafael Sacks, Ling Ma, Raz Yosef, Andre Borrmann, Simon Daum, and Uri Kattel. Semantic enrichment for building information modeling: procedure for compiling inference rules and operators for complex geometry. *Journal of Computing in Civil Engineering*, 31(6):04017062, 2017.

[304] SAE. Taxonomy and definitions for terms related to driving automation systems for on-road motor vehicles j3016_201806. https://www.sae.org/standards/content/j3016_201806, 2018. (accessed January 19, 2020).

[305] I. Salomie, T. Cioara, I. Anghel, D. Moldovan, G. Copil, and P. Plebani. An Energy Aware Context Model for Green IT Service Centers, Service-Oriented Computing. In *International Conference on Service-Oriented Computing ICSOC 2010*, volume 6568, pages 169–180. Springer, 2011.

[306] Sebastiano Sannitu. The integrated water cycle in the context of water management systems: The sardinian experience. *WATER RESOURCES FOR THE FUTURE*, page 153, 2006.

[307] Nicholas Sapankevych and Ravi Sankar. Time series prediction using support vector machines: A survey. *IEEE Computational Intelligence Magazine*, 4(2):24–38, 2009.

[308] S. Sarkar, S. Chatterjee, and S. Misra. Assessment of the suitability of fog computing in the context of Internet of Things. *IEEE Transactions on Cloud Computing*, 6(1):46–59, Jan 2015.

[309] Ina Schieferdecker, Juergen Grossmann, and Martin Schneider. Model-based security testing. *arXiv preprint arXiv:1202.6118*, 2012.

[310] Erich Schikuta, Abdelkader Magdy, and A Baith Mohamed. A framework for ontology based management of neural network as a service. In *International Conference on Neural Information Processing*, pages 236–243. Springer, 2016.

[311] Christoph Schmittner, Martin Latzenhofer, Shaaban Abdelkader Magdy, and Markus Hofer. A proposal for a comprehensive automotive cybersecurity reference architecture. In *The 7th International Conference on Advances in Vehicular Systems, Technologies and Applications*, 2018.

[312] Christoph Schmittner and Zhendong Ma. Towards a framework for alignment between automotive safety and security standards. In *International Conference on Computer Safety, Reliability, and Security*, pages 133–143. Springer, 2014.

[313] Christoph Schmittner, Zhendong Ma, and Paul Smith. Fmvea for safety and security analysis of intelligent and cooperative vehicles. In *International Conference on Computer Safety, Reliability, and Security*, pages 282–288. Springer, 2014.

[314] Christoph Schmittner, Peter Tummeltshammer, David Hofbauer, Abdelkader Magdy Shaaban, Michael Meidlinger, Markus Tauber, Arndt Bonitz, Reinhard Hametner, and Manuela Brandstetter. Threat modeling in the railway domain. In *International Conference on Reliability, Safety, and Security of Railway Systems*, pages 261–271. Springer, 2019.

[315] Erwin Schoitsch, Christoph Schmittner, Zhendong Ma, and Thomas Gruber. The need for safety and cyber-security co-engineering and standardization for highly automated

automotive vehicles. In *Advanced Microsystems for Automotive Applications 2015*, pages 251–261. Springer, 2016.

[316] Christian Schwatke, Denise Dettmering, Wolfgang Bosch, and Florian Seitz. Dahiti– an innovative approach for estimating water level time series over inland waters using multi-mission satellite altimetry. *Hydrology and Earth System Sciences 19 (10): 4345-4364*, 2015.

[317] M. Sebillo, G. Vitiello, M. Grimaldi, and A. De Piano. A citizen-centric approach for the improvement of territorial services management. *ISPRS Int. J. Geo-Information*, 9(4):223, 2020.

[318] M. Sebillo, G. Vitiello, M. Grimaldi, and D. Dello Buono. *SAFE (Safety for Families in Emergency). A citizen-centric approach for riskmanagement*, page 424–437. Springer International Publishing, 2019.

[319] M. Sebillo, G. Vitiello, M. Grimaldi, and Dimitri Dello Buono. Civil protection 2.0. a diagram-based approach for managing an emergency plan". In *22nd AGILE Conference on Geo-information Science*. Stichting AGILE.

[320] Monica Sebillo, Genny Tortora, Maurizio Tucci, Giuliana Vitiello, Athula Ginige, and Pasquale Di Giovanni. Combining personal diaries with territorial intelligence to empower diabetic patients. *Journal of Visual Languages & Computing*, 29:1–14, 2015.

[321] Ashish Sen and Muni S. Srivastava. On tests for detecting change in mean. *The Annals of Statistics*, 3(1):98–108, jan 1975.

[322] Abdelkader Magdy Shaaban, Christoph Schmittner, Gerald Quirchmayr, A Baith Mohamed, Thomas Gruber, and Erich Schikuta. Toward the ontology-based security verification and validation model for the vehicular domain. In *International Conference on Neural Information Processing*, pages 521–529. Springer, 2019.

[323] Amit Kumar Sharma and Renuka Yadav. Spam mails filtering using different classifiers with feature selection and reduction technique. In *2015 Fifth International Conference on Communication Systems and Network Technologies*, pages 1089–1093, 2015.

[324] Yuhui Shi and Russell Eberhart. A modified particle swarm optimizer. In *1998 IEEE international conference on evolutionary computation proceedings. IEEE world congress on computational intelligence (Cat. No. 98TH8360)*, pages 69–73. IEEE, 1998.

[325] Yuhui Shi and Russell C Eberhart. Fuzzy adaptive particle swarm optimization. In *Proceedings of the 2001 congress on evolutionary computation (IEEE Cat. No. 01TH8546)*, volume 1, pages 101–106. IEEE, 2001.

[326] Robert H. Shumway and David S. Stoffer. *Time Series Analysis and Its Applications: With R Examples*. Springer, 2017.

[327] Jack Sieber. Weap water evaluation and planning system. In *3rd International Congress on Environmental*, Burlington, Vermont, USA, 2006.

[328] M. Silva, D. Freitas, E. Neto, C. Lins, V. Teichrieb, and J. M. Teixeira. Glassist: Using augmented reality on google glass as an aid to classroom management. In *2014 XVI Symposium on Virtual and Augmented Reality*, pages 37–44, May 2014.

[329] Smart city cluster Task 3. Website: `http://ec.europa.eu/newsroom/dae/document.cfm?doc_id=8011`.

[330] J. Smith and J.U.M. Smith. *Computer Simulation Models*. Techniques-case studies. Hafner Publishing Company, 1968.

[331] Chaoming Song, Zehui Qu, Nicholas Blumm, and Albert-László Barabási. Limits of predictability in human mobility. *Science*, 327(5968):1018–1021, 2010.

[332] Cagatay Sonmez, Atay Ozgovde, and Cem Ersoy. Edgecloudsim: An environment for performance evaluation of edge computing systems. *Transactions on Emerging Telecommunications Technologies*, 29(11):e3493, 2018.

[333] SPARQL specification. Website: `http://www.w3.org/TR/rdf-sparql-query/`.

[334] SWRL specification. Website: `http://www.w3.org/Submission/SWRL/`.

[335] Michiel Stas, Jos Van Orshoven, Qinghan Dong, Stien Heremans, and Beier Zhang. A comparison of machine learning algorithms for regional wheat yield prediction using ndvi time series of spot-vgt. In *2016 Fifth International Conference on Agro-Geoinformatics (Agro-Geoinformatics)*, pages 1–5, 2016.

[336] Vladeta Stojanovic, Matthias Trapp, Rico Richter, Benjamin Hagedorn, and Jürgen Döllner. Towards the generation of digital twins for facility management based on 3d point clouds. In *Proceeding of the 34th Annual ARCOM Conference*, pages 270–279, 2018.

[337] Christof Strauch. *NoSQL databases*. Lecture Notes, Stuttgart Media University, 2011.

[338] Autodesk Dynamo Studio. Dynamo primer for v1. 3, 2017.

[339] A. Sumida, T. Miyaura, and H. Torii. Relationships of tree height and diameter at breast height revisited: analyses of stem growth using 20-year data of an even-aged chamaecyparis obtusa stand. *Tree Physiology*, 33(1):106–118, 2013.

[340] Synopsys. Dude, where's my autonomous car? the 6 levels of vehicle autonomy. https://www.synopsys.com/automotive/autonomous-driving-levels.html, 2019. (accessed January 27, 2020).

[341] SWINDON Silicon Systems. Determining functional safety levels for automotive applications. https://www.swindonsilicon.com/determining-functional-safety-levels-for-automotive-applications/, 2019. (accessed January 25, 2020).

[342] I. Salomie T. Cioara, I. Anghel. Methodology for Energy Aware Adaptive Management of Virtualized Data Centers. In *Energy Efficiency*, page 475–498, 2017.

[343] T. Taleb, K. Samdanis, B. Mada, H. Flinck, S. Dutta, and D. Sabella. On multi-access edge computing: A survey of the emerging 5G network edge cloud architecture and orchestration. *IEEE Communications Surveys Tutorials*, 19(3):1657–1681, Third Quarter 2017.

[344] Lei Tang, Lin Li, Shen Ying, and Yuan Lei. A full level-of-detail specification for 3d building models combining indoor and outdoor scenes. *ISPRS International Journal of Geo-Information*, 7(11):419, 2018.

[345] Qinghui Tang, Tridib Mukherjee, Sandeep KS Gupta, and Phil Cayton. Sensor-based fast thermal evaluation model for energy efficient high-performance datacenters. In *2006 Fourth international conference on intelligent sensing and information processing*, pages 203–208. IEEE, 2006.

[346] Shu Tang, Dennis R Shelden, Charles M Eastman, Pardis Pishdad-Bozorgi, and Xinghua Gao. A review of building information modeling (bim) and the internet of things (iot) devices integration: Present status and future trends. *Automation in Construction*, 101:127–139, 2019.

[347] Apache HBase Team. *Apache HBase Reference Guide v3*. Apache, 2019.

[348] A Teruzzi, P Di Cerbo, G Cossarini, E Pascolo, and S Salon. Parallel implementation of a data assimilation scheme for operational oceanography: The case of the medbfm model system. *Computers & Geosciences*, 124:103–114, 2019.

[349] Channary Thay, Vasaka Visoottiviseth, and Sophon Mongkolluksamee. P2p traffic classification for residential network. In *2015 International Computer Science and Engineering Conference (ICSEC)*, pages 1–6, 2015.

[350] Chris D. Thomas, Michael C. Singer, and David A. Boughton. Catastrophic extinction of population sources in a butterfly metapopulation. *The American Naturalist*, 148(6):957–975, 1996.

[351] J. J. Thomas and K. A. Cook. *Illuminating the Path*. IEEE Computer Society.

[352] J. J. Thomas and K. A. Cook. A visual analytics agenda. *IEEE Computer Graphics and Applications*, 26(1):10–13, 1 2006.

[353] S. Thrun. Efficient exploration in reinforcement learning. *Technical Report CMU-CS-92-102. Pittsburgh, PA: Carnegie Mellon University*, pages 92–102, 1992.

[354] Timescale. Timescaledb documentation v1.7 https://docs.timescale.com/latest.

[355] Mattia Tomasoni, Andrea Capponi, Claudio Fiandrino, Dzmitry Kliazovich, Fabrizio Granelli, and Pascal Bouvry. Why energy matters? profiling energy consumption of mobile crowdsensing data collection frameworks. *Pervasive and Mobile Computing*, 51:193 – 208, 2018.

[356] Mohan Manubhai Trivedi, Tarak Gandhi, and Joel McCall. Looking-in and looking-out of a vehicle: Computer-vision-based enhanced vehicle safety. *IEEE Transactions on Intelligent Transportation Systems*, 8(1):108–120, 2007.

[357] Charles Truong, Laurent Oudre, and Nicolas Vayatis. Selective review of offline change point detection methods. *Signal Processing*, 167:107299, 2020.

[358] Can Tunca, Nezihe Pehlivan, Nağme Ak, Bert Arnrich, Gülüstü Salur, and Cem Ersoy. Inertial sensor-based robust gait analysis in non-hospital settings for neurological disorders. *Sensors*, 17(4):825, 2017.

[359] Shane R Turner, Todd E Erickson, Miriam Rojas, and David J Merritt. The restoration seed bank initiative– a focus on biodiverse restoration in the semi-arid pilbara of western australia. *BGjournal*, 13(2):20–23, 2016.

[360] C. B. T.W. Rondeau. *Artificial Intelligence in Wireless Communications*. Boston, USA: Artech House, 2009.

[361] Raihan Ul Islam, Karl Andersson, and Mohammad Shahadat Hossain. A web based belief rule based expert system to predict flood. In *Proceedings of the 17th International conference on information integration and web-based applications & services*, page 3. ACM, 2015.

[362] R. Urgaonkar, B. Urgaonkar, M. Neely, and A. Sivasubramaniam. Optimal power cost management using stored energy in data centers. In *SIGMETRICS '11: Proceedings of the ACM SIGMETRICS Joint International Conference on Measurement and Modeling of Computer Systems*, page 221–232. ACM, 2011.

[363] Gerard van der Schrier, J Barichivich, KR Briffa, and PD Jones. A scpdsi-based global data set of dry and wet spells for 1901–2009. *Journal of Geophysical Research: Atmospheres*, 118(10):4025–4048, 2013.

[364] Gokula Vijaykumar Annamalai Vasantha, Rajkumar Roy, Alan Lelah, and Daniel Brissaud. A review of product–service systems design methodologies. *Journal of Engineering Design*, 23(9):635–659, 2012.

[365] Piergiorgio Vitello, Andrea Capponi, Claudio Fiandrino, Paolo Giaccone, Dzmitry Kliazovich, and Pascal Bouvry. High-precision design of pedestrian mobility for smart city simulators. In *IEEE International Conference on Communications (ICC)*, pages 1–6, May 2018.

[366] X. Wang W. Zheng, K. Ma. Exploiting Thermal and Energy Storage to Cut the Electricity Bill for Datacenter Cooling. In *Proceedings of the 2012 8th International Conference on Network and Service mManagement (CNSN)*, pages 9–27. IEEE, 2012.

[367] W3C. Vocabularies. https://www.w3.org/standards/semanticweb/ontology, 2019.

[368] Andreas Wagner, William O'Brien, and Bing Dong. Exploring occupant behavior in buildings. *Wagner, A., O'Brien, W., Dong, B., Eds*, 2018.

[369] S. Wang, G. Ding, and M. Zhong. On spatial data mining under big data. *Journal of China Academy of Electronics and Information Technology*, 8:8–17, 2013.

[370] S. Wang and H. Yuan. Spatial data mining in the context of big data. In *19th IEEE International Conference on Parallel and Distributed Systems, ICPADS 2013, Seoul, Korea, December 15-18, 2013*, pages 486–491. IEEE Computer Society, 2013.

[371] S. Wang and H. Yuan. Spatial data mining: A perspective of big data. *International Journal of Data Warehousing and Mining*, 10(4):50–70, 10 2014.

[372] Ying-Ming Wang, Jian-Bo Yang, and Dong-Ling Xu. Environmental impact assessment using the evidential reasoning approach. *European Journal of Operational Research*, 174(3):1885–1913, 2006.

[373] C Wanigasooriya, Malka N Halgamuge, and Azeem Mohammad. The analysis of anticancer drug sensitivity of lung cancer cell lines by using machine learning clustering techniques. *International Journal of Advanced Computer Science and Applications*, 8(9), 2017.

[374] Nathan Wells, Steve Goddard, and Michael J Hayes. A self-calibrating palmer drought severity index. *Journal of Climate*, 17(12):2335–2351, 2004.

[375] Philip West. *Tree and Forest Measurement*. Springer-Verlag Berlin Heidelberg, 2009.

[376] P. Wette, M. Dräxler, and A. Schwabe. MaxiNet: Distributed emulation of software-defined networks. In *IFIP Networking*, pages 1–9, 2014.

[377] F Woodward, G Fogg, and U Heber. The impact of low temperatures in controlling the geographical distribution of plants. *Philosophical Transactions of the Royal Society of London. B, Biological Sciences*, 326(1237):585–593, 1990.

[378] Pin Wu, Junwu Sun, Xuting Chang, Wenjie Zhang, Rossella Arcucci, Yike Guo, and Christopher C Pain. Data-driven reduced order model with temporal convolutional neural network. *Computer Methods in Applied Mechanics and Engineering*, 360:112766, 2020.

[379] D Xiao, CE Heaney, F Fang, L Mottet, R Hu, DA Bistrian, E Aristodemou, IM Navon, and CC Pain. A domain decomposition non-intrusive reduced order model for turbulent flows. *Computers & Fluids*, 2019.

[380] Qi Xiong, Guiming Chen, Zhaojun Mao, Tianjun Liao, and Leilei Chang. Computational requirments analysis on the conjunctive and disjunctive assumptions for the belief rule base. In *2017 International Conference on Machine Learning and Cybernetics (ICMLC)*, volume 1, pages 236–240. IEEE, 2017.

[381] J. Xu, B. Palanisamy, H. Ludwig, and Q. Wang. Zenith: Utility-aware resource allocation for edge computing. In *Proc. of IEEE EDGE*, pages 47–54, Jun 2017.

[382] X. Wang Y. Zhang, Y. Wang. Exploiting Thermal Energy Storage to Reduce Data Center Capital and Operating Expenses. In *2014 IEEE 20th International Symposium on High Performance Computer Architecture (HPCA)*, pages 132–141. IEEE, 2014.

[383] Da Yan, William O'Brien, Tianzhen Hong, Xiaohang Feng, H Burak Gunay, Farhang Tahmasebi, and Ardeshir Mahdavi. Occupant behavior modeling for building performance simulation: Current state and future challenges. *Energy and Buildings*, 107:264–278, 2015.

[384] Zhixian Yan. *Semantic trajectories: computing and understanding mobility data*. PhD thesis, Verlag nicht ermittelbar, 2011.

[385] Zhixian Yan, Dipanjan Chakraborty, Christine Parent, Stefano Spaccapietra, and Karl Aberer. Semantic trajectories: Mobility data computation and annotation. *ACM Transactions on Intelligent Systems and Technology (TIST)*, 4(3):1–38, 2013.

[386] Jian-Bo Yang, Jun Liu, Jin Wang, How-Sing Sii, and Hong-Wei Wang. Belief rule-base inference methodology using the evidential reasoning approach-rimer. *IEEE Transactions on systems, Man, and Cybernetics-part A: Systems and Humans*, 36(2):266–285, 2006.

[387] Jian-Bo Yang and Pratyush Sen. A general multi-level evaluation process for hybrid madm with uncertainty. *IEEE Transactions on Systems, Man, and Cybernetics*, 24(10):1458–1473, 1994.

[388] Wu Yanyan and Zhou Guobing. The analysis of electronic commerce project risk with the bayes network. In *Proceedings of 2011 International Conference on Electronic & Mechanical Engineering and Information Technology*, pages 3650–3653, 2011.

[389] Y. Yao, L. Huang, A. Sharma, L. Golubchik, and M. Neely. Data Centers Power Reduction: A Two Time Scale Approach for Delay Tolerant Workloads. In *2012 Proceedings IEEE INFOCOM*, pages 431–1439. IEEE, 2012.

[390] Wonjae Yoo, Hyoungsub Kim, and Minjae Shin. Stations-oriented indoor localization (soil): A bim-based occupancy schedule modeling system. *Building and Environment*, 168:106520, 2020.

[391] Wucherl Yoo, Alex Sim, and Kesheng Wu. Machine learning based job status prediction in scientific clusters. In *2016 SAI Computing Conference (SAI)*, pages 44–53, 2016.

[392] Y. Zeng, M. Chao, and R. Stoleru. EmuEdge: A hybrid emulator for reproducible and realistic edge computing experiments. In *Proc. IEEE ICFC*, pages 153–164, June 2019.

[393] Zhi-Hui Zhan, Jun Zhang, Yun Li, and Henry Shu-Hung Chung. Adaptive particle swarm optimization. *IEEE Transactions on Systems, Man, and Cybernetics, Part B (Cybernetics)*, 39(6):1362–1381, 2009.

[394] Juan Zhang and Xin-yuan Huang. Measuring method of tree height based on digital image processing technology. In *2009 First International Conference on Information Science and Engineering*, pages 1327–1331, 2009.

[395] Jie Zhao, Bertrand Lasternas, Khee Poh Lam, Ray Yun, and Vivian Loftness. Occupant behavior and schedule modeling for building energy simulation through office appliance power consumption data mining. *Energy and Buildings*, 82:341–355, 2014.

[396] Y Zheng. Trajectory data mining: An overview. acm transaction on intelligent systems and technology. *ACM Transactions on Intelligent Systems and Technology*, 6(3), 2015.

[397] Yuren Zhou, Billy Pik Lik Lau, Chau Yuen, Bige Tunçer, and Erik Wilhelm. Understanding urban human mobility through crowdsensed data. *IEEE Communications Magazine*, 56(11):52–59, Nov 2018.

[398] Zhi-Jie Zhou, Guan-Yu Hu, Changhua hu, and Cheng-Lin Wen. A survey of belief rule-base expert system. *IEEE Transactions on Systems, Man, and Cybernetics: Systems*, PP:1–15, 11 2019.

[399] Jiangcheng Zhu, Shuang Hu, Rossella Arcucci, Chao Xu, Jihong Zhu, and Yi-ke Guo. Model error correction in data assimilation by integrating neural networks. *Big Data Mining and Analytics*, 2(2):83–91, 2019.

[400] Changliang Zou, Guosheng Yin, Long Feng, Zhaojun Wang, et al. Nonparametric maximum likelihood approach to multiple change-point problems. *The Annals of Statistics*, 42(3):970–1002, 2014.

Index